The Future of the World's Forests

WORLD FORESTS

Series Editors

MATTI PALO
PhD, Independent Scientist, Finland, Affiliated Professor CATIE, Costa Rica

JUSSI UUSIVUORI
Finnish Forest Research Institute METLA, Finland

Advisory Board

Janaki Alavalapati, University of Florida, USA
Joseph Buongiorno, University of Wisconsin, USA
Jose Campos, CATIE, Costa Rica
Sashi Kant, University of Toronto, Canada
Maxim Lobovikov, FAO/Forestry Department, Rome
Misa Masuda, University of Tsukuba
Roger Sedjo, Resources for the Future, USA
Brent Sohngen, Ohio State University, USA
Yaoqi Zhang, Auburn University, USA

World Forests Description

As forests stay high on the global political agenda, and forest-related industries diversify, cutting edge research into the issues facing forests has become more and more transdisciplinary. With this is mind, Springer's World Forests series has been established to provide a key forum for research-based syntheses of globally relevant issues on the interrelations between forests, society and the environment.

The series is intended for a wide range of readers including national and international entities concerned with forest, environmental and related policy issues; advanced students and researchers; business professionals, non-governmental organizations and the environmental and economic media.

Volumes published in the series will include both multidisciplinary studies with a broad range of coverage, as well as more focused in-depth analyses of a particular issue in the forest and related sectors. Themes range from globalization processes and international policies to comparative analyses of regions and countries.

Jim Douglas · Markku Simula

The Future of the World's Forests

Ideas vs Ideologies

Jim Douglas
Australian National University
College of Medicine, Biology and Environment
Fenner School of Environment and Society
Canberra, ACT
Australia
jim.douglas@anu.edu.au

Markku Simula
University of Helsinki
Department of Agriculture and Forestry
Finland
markku.simula@ardot.fi

ISBN 978-90-481-9581-7   e-ISBN 978-90-481-9582-4
DOI 10.1007/978-90-481-9582-4
Springer Dordrecht Heidelberg London New York

Library of Congress Control Number: 2010935295

© Springer Science+Business Media B.V. 2010
No part of this work may be reproduced, stored in a retrieval system, or transmitted in any form or by any means, electronic, mechanical, photocopying, microfilming, recording or otherwise, without written permission from the Publisher, with the exception of any material supplied specifically for the purpose of being entered and executed on a computer system, for exclusive use by the purchaser of the work.

Cover illustration: Cleared forest for agriculture in southern Mexico (2005). Photograph by Markku Simula.

Printed on acid-free paper

Springer is part of Springer Science+Business Media (www.springer.com)

We dedicate this book to the memory of Alf Leslie, whom we lost in January 2009. Over six or more decades of engagement in forests, especially the tropical forests, Alf was one of the most incisive and original thinkers about how we should approach forests, and one of the most generous of advisors and colleagues. It would be fair to say that much of what has been said and written about the tropical forests over the last three decades reflects – consciously or otherwise – significant elements that Alf, and his great partner Jack Westoby, laid down years before, and this book is no exception.

To Alf: iconoclast, mentor, and friend.
JD and MS

Acknowledgements

In the course of more than half a lifetime spent in a particular field – in our case, the large and by no means level one of international forestry- one incurs debts of knowledge, insight, cooperation, willingness to apply forensic questioning to ideas and ideologies, owed to a huge number of people. We have encountered them in the many international development agencies, multilateral development banks, government agencies, research and teaching institutions, non-government organizations and other civil society groups we have worked for, or with. We have learned much from many who have trenchantly disagreed with us and, whether they like it or not, we wish to thank them for their engagement with us anyway. We have been fortunate to have been able to travel widely in the forests of the developing world, and to meet many of the people who live and work in them, and depend upon them for their livelihood. We hope that some of what appears in this book will give due voice to at least some of the concerns and priorities of these people.

Special thanks are due here to our publisher, Springer, in particular to Catherine Cotton and Ria Kanters, for patience and forbearance during the long birthing process of this book, and to reviewers for their suggestions and help. We are grateful to The Prince's Rainforests Project for their kind permission to use material in this book done under their auspices, and to the World Bank in similar vein.

Our partners, Silvia Xavier and Nicole, have been willing to provide feedback and reactions to what appears here, and, of course, have also travelled with us on this long journey, and managed somehow to put up with us for the duration. For their support, and love, we are deeply grateful.

Contents

Part I Issues and Questions

1 Disappearing Rainforests: New Solutions 3
   1.1 Introduction 3
   1.2 But Why Another Book on Forests? 5
        1.2.1 There Is Certainly no Shortage of Material 5
        1.2.2 There Are Still Some Things Left to Say 8
   1.3 The Dynamics of Forest Loss 9
        1.3.1 Natural Forests Are in Decline 9
        1.3.2 So, Is the World Running Out of Wood? 10
        1.3.3 Can Developing International Trade and Natural
               Forests Survival Be Compatible? 11
        1.3.4 What Else Is Happening to Natural Forests? 12
   1.4 Setting the Scene for Sustainability: Valuation;
        and Financing of the Forests 14
   1.5 What Has Happened to Forests Sustainability So Far? 15
        1.5.1 Valuing Forests: The Full Ecosystem;
               and the Carbon Values 15
        1.5.2 Financing Change in the Forests 16
   References 17

2 Global Forests: Debate and Dysfunction 19
   2.1 Defining the Problem #1: A Dysfunctional Dialogue 20
        2.1.1 The Problem You See Depends on Where You Stand 22
        2.1.2 The Search for Perspective: "You Talk; We Chop" 23
        2.1.3 Re-evaluation, or Late Onset Apostasy? 25
        2.1.4 Re-focusing the Discussion 26
   2.2 Defining the Problem #2: Sustainability and Forests
        Value – The Basic Issues 27
        2.2.1 What Have Forests To Do with Global Economics? 28
        2.2.2 A Question of Value 29
        2.2.3 Sustainability: The Elusive Objective 29

		2.2.4	Why Has Sustainable Forest Management Not Worked?	30
		2.2.5	Never Mind the Differences on Sustainability: Can't We All Just Get Along?	33
		2.2.6	Asking the Wrong Questions	33
		2.2.7	The Right Question	35
		2.2.8	Implications for Some Popular Current Approaches to Sustainability	35
	2.3	The Global Dialogue on Forests: Moribunds, Mercantilists, and Manicheans		36
	References			39
3	**The State of Global Forest Resources**			41
	3.1	The State of the World's Forests		43
		3.1.1	The Historical Picture	43
		3.1.2	Global Forest Cover and Cover Change	44
	3.2	Tropical Rainforests: A Key Concern		48
		3.2.1	Forest Ecosystem Services	49
		3.2.2	Irreversibility of Rainforest Loss: A Key Concern	51
	3.3	The Implications of Reducing Deforestation		51
	References			52

Part II The Dynamics of Forest Loss

4	**Are Trade and Forests Survival Compatible?**			55
	4.1	Where Trade Is Going: Emerging Trends		56
	4.2	How Future Demand Can Be Met: Rapidly Changing Supply Patterns		58
		4.2.1	The Raw Material Base	58
		4.2.2	Forest Industry	59
	4.3	Are Impacts of Trade Liberalization on Forests Positive?		60
		4.3.1	Winners and Losers Under Trade Liberalization	61
	4.4	Can Trade Rules Differentiate Sustainably Produced Forest Products?		62
		4.4.1	Protection of Forest-Related Intellectual Property Rights	63
		4.4.2	Trade in Intergovernmental Forest Agreements	64
		4.4.3	Pressures from Non-governmental Organizations	65
		4.4.4	Taking Stock	66
	4.5	Has Certification of Forest Management Created Value for Forest Resources?		67
		4.5.1	Lagging Developing Countries and Uncertain Market Benefits	67
		4.5.2	A Tug-of-War Between International Schemes	68
		4.5.3	Improving Effectiveness of Certification	70

4.6	A Distorted Playing Field: Addressing Illegal Logging		71
	4.6.1	Causes of Illegal Logging and trade	72
	4.6.2	Emerging Market Requirements	73
	4.6.3	Trade Measures in Combating Illegal Logging and Trade	74
	4.6.4	Can Certification Impact upon Illegal Logging?	75
	4.6.5	Taking Stock	76
4.7	New Opportunities and Challenges for Trade in the Valuation of Forests		76
4.8	Conclusions		78
References			79

5 Deforestation: Causes and Symptoms ... 81

5.1	Rainforests: A Tragedy of the Commons?		82
5.2	Agricultural Technology and Deforestation		84
5.3	The Impact of Burgeoning Plantation and Grazing Commodities		85
	5.3.1	Cattle Ranching (Brazil)	86
	5.3.2	Soy	87
	5.3.3	Biofuels	88
	5.3.4	Palm Oil	90
	5.3.5	Forest Industry Plantations and the Pulp and Paper Sector	91
5.4	Some Other Factors in Deforestation		94
	5.4.1	Mining	94
	5.4.2	Wood Fuel	95
	5.4.3	Pioneer Shifting Cultivation	95
	5.4.4	Infrastructure	96
5.5	Illegal Logging		96
5.6	Commentary on Some Corrective Options		98
	5.6.1	International Demand Management	98
	5.6.2	Eliminating Perverse Production Incentives	100
5.7	Separating Causes and Symptoms		101
References			103

Part III Sustainability and Valuation of the Forests

6 Sustainability Versus Ideology in the Forests ... 107

6.1	Global Environmental Sustainability: The Shifting Paradigm		108
	6.1.1	Malthus: The Original Prophet of Economic Doom	108
	6.1.2	The Inheritors of Malthus	109
	6.1.3	Growth Protagonists Push Back	110
	6.1.4	The Environmental Kuznets Curve	111
	6.1.5	Taking Stock	112

6.2	Forests and the Broader Economy	114
	6.2.1	Applying the Environmental Kuznets Curve to Forests	114
	6.2.2	Forests in the Broader Economy	115
	6.2.3	Financing Sustainability in Forests Has Been Inadequate	117
6.3	Multilateral Agreements on Global Environmental Sustainability	118
	6.3.1	The Stockholm Agreement	119
	6.3.2	The Brundtland Report	119
	6.3.3	The Rio Earth Summit	120
	6.3.4	The Kyoto Protocol	121
	6.3.5	The World Summit on Sustainable Development	121
6.4	A Brief Look at Multilateral Involvement in Forests	122
	6.4.1	Sustainable Forest Management	123
	6.4.2	The Tropical Forestry Action Plan	124
	6.4.3	The International Tropical Timber Organization	127
	6.4.4	The Intergovernmental Dialogue on Forests	127
	6.4.5	The Forest Law Enforcement and Governance Initiatives	130
6.5	Forest Policy in the World Bank: Ideas vs Ideologies	130
	6.5.1	The "Chilling Effect" of Bank Forests Sector Policy	131
	6.5.2	The New World Bank Forests Sector Strategy and Policy	133
	6.5.3	The Chill Is (Not) Gone	136
	6.5.4	Problems with Ideology: The Conservation International case	138
6.6	Developing Perspectives on Sustaining Forests	142
References	144

7 Financing Forests Sustainability from Ecosystem Values ... 147

7.1	The Failure of Forests Sustainability: A Question of Perceived Value	148
	7.1.1	Valuing the Natural Forests: Qualitative Assessments	149
	7.1.2	Quantifying Ecosystem Values	150
	7.1.3	Forests and Climate Change	151
7.2	Stored Forest Carbon: Leading the New Sustainability Paradigm	156
	7.2.1	Will Reducing Tropical Rainforest Deforestation Be a Cost Effective GHG Strategy?	156
	7.2.2	A Case Study: Oil Palm in Indonesia	159
7.3	Would Rainforest Governments Finance Sustainability in Forests for Carbon?	165
	7.3.1	Forest Loss and Broad Economic Change	166

	7.4	Financing Reduced Emissions from Deforestation and Forest Degradation	171
		7.4.1 REDD Has Been a Long Time Coming	172
		7.4.2 The Basics of REDD	175
		7.4.3 Issues and Differences for Consideration Under REDD	176
		7.4.4 Some Controversies and the "Agenda Loading" Issue	182
	7.5	Investing in Reduced Deforestation Ahead of REDD	186
		7.5.1 An Emergency Package for Tropical Forests	187
	References		191
8	**Final Thoughts**		195
	8.1	The Search for a New Paradigm	196
	8.2	What Does All This Mean for the Forests?	197
		8.2.1 Formulating the Link Between Forests and Capital	199
		8.2.2 Balancing Equity and Effectiveness	200
		8.2.3 Keeping Watch Over the Market Instrument	201
		8.2.4 International Debates on Forests: Impotency and Intransigence	204
	8.3	Some Final Words	206
	References		207
Index			209

Part I
Issues and Questions

Chapter 1
Disappearing Rainforests: New Solutions

> *Why should men quarrel here, where all possess*
> *As much as they can hope for by success?*
> *None can have most where nature is so kind*
> *As to exceed man's use, though not his mind*
>
> Sir Robert Howard (1626–1698) and John Dryden (1631–1700)

Abstract In this chapter, the basic motivations for addressing the global deforestation issue – what the major issues and constraints involved will be; and how these might be dealt with – are laid out. The (possibly contentious) observation is made that while the emotion that loss of the large, charismatic natural forests of the developing world generates in many observers of the forest scene worldwide is potentially a driving force for change, it is also prone to misuse: concerned individuals can be led into extreme or unrealistically ideological positions. It is argued that, like it or not, some recognition of legitimate commercial use of forests, and of the importance of national sovereignty in how this is done are as necessary, in the solution of deforestation, as are concerns about loss of biodiversity, unacceptable treatment of local communities in forest areas, and other issues. It is further argued that the large questions as to how global forests can be saved and wisely used must be addressed directly, and in an integrated manner – something which has not happened to any great extent to date: how much forest can and should be sustained? Who should do this? Who will pay? Who should benefit most?

1.1 Introduction

Since you have opened this book, we are assuming there is something about forests, on the grand global scale, that interests you. We would be prepared to go further, and surmise that, unless you are a recent arrival on this planet, it is likely you have picked up on a feeling that something is badly wrong in the forests, and that, so far, no effective, large solutions have been developed: man's mind, it seems, has led us to a situation where his use has indeed begun to outpace nature's kindness.

If this is your feeling, then you share it with many in what might loosely be termed the international forests community: a broad amalgam of government officials in forested countries (and, for that matter, in countries whose forests are depleted or naturally poor and inadequate to the country's needs), people from development assistance agencies, multilateral development banks, environmental and social non-governmental organizations, some of the more progressive elements of forest industries, academics, researchers, and a large number of other interest groups. It would be fair to say that the mood on forests in this community is predominantly one of unease and uncertainty, ranging through to profound gloom in some cases.

One of the reasons for this, it must be said, is that natural forests are charismatic: Relatively untouched natural forests – especially in the tropics and the boreal zone – are one of the last great natural resource frontiers left on the planet: the oceans are the other. Many observers and practitioners involved in forests view the stakes as being very high for this resource, and the risks to it as being extremely disturbing. In a sense, forests are seen by many as the whales of the land: beautiful, primeval; immensely precious. They are accorded sobriquets such as "old growth" and "ancient" (even in many cases where they are neither) to add to their charisma.

It is not difficult to understand the emotional chord that is plucked when images are shown of great trees hitting the ground in a logging operation, or ablaze in fires of deliberate, human origin. If you actually visit a tropical rainforest – just as if you gain a close-up view of whales at sea – your sense of their value, their majesty, will if anything be enhanced. From the moist, cool shade of the forest floor, you will gaze up at the high primary canopy vaulted like a cathedral roof over the complex layers of understorey; you may sit quietly for a period, and feel the enormous hum of life all about you. Whether you can see them or not, you will know you are probably in close proximity to more species of your fellow inhabitants of this planet than at virtually any other place on it that you might visit. You may even experience a feeling that you have returned to your own biological origins – we are, after all, by some accounts a tropical species ourselves.

If you have been told, before your visit, that sites such as these are a vital element in maintaining a whole suite of natural resources – water catchments and land stability; coastal marine systems; a broad swathe of global biodiversity; and the balance of the earth's climate itself – you will believe it: the sheer biological power of this place can be felt as you stand in it.

Even if we change the forest scene here to the austere, icy beauty of a boreal forest in the north of Russia or Canada, or a stand of giant eucalypts in the temperate southern island state of Australia – Tasmania – it is likely you will still experience similar feelings. When you are there, the idea that this very piece of forest you are in might soon be logged over, or simply burnt down and replaced by some other form of land use, will seem deeply unacceptable, and it is easy for feelings of concern to overflow into outrage.

This emotional reaction is one of the best instruments we have for raising consciousness on the need for protection of forests; but at the same time, if misused, it can be the major impediment to building a realistic consensus for getting the job done in the field. Passion to get something done, something changed, is necessary, but we must all face the reality that if we react purely, or primarily, from emotion, or a sort of ecological Gnosticism, requiring practitioners to acquire a transcendent spiritual knowledge (and there are many groups interested in forests who will want us to do precisely this), then we are increasing the risk that we will lose the real argument over protecting and sustaining natural forests. As we will see time and again through this book, we need to differentiate carefully between arguments we must win, and those we would simply like to win.

The point of differentiation arises from the reality that, to get some positive result from a discussion of what to do with these forests, we need to be as sympathetic to issues of local livelihood, commercial activity, and national sovereignty in these forests, as to the need to protect them and prevent their degradation and destruction. None of these imperatives can be subjugated to a strong wish on the part of many people – especially those who live outside those developing countries with large endowments of natural forests – to move the balance of management of these forests more towards pure protection – or sometimes even sustainable management – of significant areas of them, unless a great deal more of substance is brought to the negotiating table to bring this about. What that substance might be, and how it might be deployed to the task of improving the situation, will be a major concern of this book.

As will by now be clear, our primary interest in this book is the survival and sustainability of the very large areas of natural forests remaining in developing countries. We will focus considerable attention on tropical rainforests, on the grounds that these are the most demanding of global concern. This is not to suggest that we are uninterested in forests in developed countries, or those in emerging economies of eastern Europe and Russia. We do not wish to infer that these forests do not present their own range of significant management and conservation issues. We do suggest, however, that the financial and technical options for resolving these issues are much more within the grasp of those forest-owning countries themselves than is the case for the forests of south and central America, Africa, the South Asia sub-continent, South-East Asia and the Pacific islands region.

Our view is that if the natural forests issue as it plays out in these regions of the world is allowed to become (some would say *remain*) a fringe issue on the international stage, then the cause of global forests and the potential benefits they bring to the globe will be lost, no matter what happens elsewhere. This will be so no matter how passionate, dedicated and committed the fringe dwellers of the forests issue may be. This is the reason why, amongst those we hope will read this book, will be the person who is not necessarily specialized and working in some aspect of forests, environment and conservation, but who nevertheless recognizes that managing the world's natural forests for a better outcome than is currently indicated, is important for all of us.

1.2   But Why Another Book on Forests?

1.2.1   *There Is Certainly no Shortage of Material*

Readers familiar with the international forestry scene will be aware that a fairly cursory trawl with any general web search engine, accessible academic and research literature sources, and the available grey literature from a large and varied collection of interest groups and organizations, would yield a collection of paper on the general subject of global forests that would test the load limits of a good-sized logging truck.

Given this, we had to acknowledge when considering this project that we could run the risk of testing the forbearance of some readers – but we do hope to show that there is more of interest to be said on this subject. Obviously, we ride on the backs of much of that earlier work, and we provide the briefest sampling below of the sort of book length material that has become available; much more of this will be cited as we proceed with our arguments in this book.

David Humphreys published a book on forest politics in the mid-1990s (Humphreys 1996), not long after the enthusiasms generated by the Rio Earth Summit began to be replaced with the more disappointing realities of what the global community was prepared to venture in support of conserving and sustaining the world's great forests. The book leads with the central political issue of deforestation, and then proceeds with a critical examination of the various multilateral treaties and programmes on international forestry which have been introduced from the 1980s onwards: the Tropical Forestry Action Programme; the International Tropical Timber Organization; forest negotiations under the UNCED process; and others. Humphreys concludes that governments of the North cannot expect developing countries with large forest resources to yield sovereignty without some of their concerns being addressed, while on the other hand, those developing countries are unlikely to extract concessions from the North unless serious commitments to conservation are made. He raises the importance of involvement of communities living in or near forests in any workable solution, suggesting that recognition of rights of these people is not just an environmental imperative, but is also one of social justice.

In his second book on this broad subject (Humphreys 2006) extends his examination of international initiatives on forests, reviewing the Intergovernmental Panel on Forests, and its successors the Intergovernmental Forum on Forests, and the present United Nations Forum on Forests, as well as the World Bank's 2003 forests sector strategy (World Bank 2004), and other programmes. Humphreys develops his general proposition further by arguing for a democratization of global governance and a fundamental restructuring of the regulatory environment in the forests sector so that final decision making authority is restored to the local level.

The community involvement theme has featured strongly in international literature on forests for the past decade in particular. Jerry Vanclay, Ravi Prabhu and Fergus Sinclair in their book (Vanclay et al. 2006) provide guidelines for communities to develop and manage their natural resources (including forests) via participatory modelling, and provide a wide range of field examples of its application.

Moria Moelino, Eva Wollenberg and Godwin Limberg have edited a collection of studies (Moelino et al. 2008) by a team of researchers in Malinau, Indonesian Borneo, one of the world's richest forest areas. The book discusses the theoretical framework for devolution of agency over forests to local level, the progress of the devolution process in the area studies, and the broad issues of property rights, governance and the politics of the process. The book argues that cultural alliances, especially among minority groups, are taking greater prominence in the determination of forest policy. Importantly, it draws out the salient points for other international contexts including the important determination that cultural alliances, especially among ethnic minorities, are taking on greater prominence and ethnic minorities are finding new ways to influence forest policy in the world's richest forests.

Another theme which has been developing alongside that of local governance in forests is combining the approaches and systems which have been developed for ecosystem management in recent years, with those designed for sustainable forest management, leading to an overall landscape approach to the problem. In the past, there has been a tendency to see these as different methods to be applied for different purposes, in different forests. Jeff Sayer and Stewart Maginnis have edited a collection of case studies (Sayer and Maginnis (eds) 2007) which have yielded innovative ways of combining the two approaches, providing insights into how criteria and indicators for managing forests areas have been used (and misused) in the past, and lessons to be drawn from the field experience with new, more integrated approaches.

It is important to bear in mind, when considering these issues of governance and management of forests, that markets for what forests produce are, and will remain, an important element in how they are used and managed. Matti Palo, Jussi Uusivuori, and Gerardo Mery have in their book (Palo et al. 2001) taken a closer look at the role of markets and policies in transferring the value of forests to societies. They acknowledge that there are large externalities present in determining what happens to forests, and that these may be well beyond the reach of sectoral markets to influence; we will address this issue in this book, and it is also implicit in the discussion in the following paragraph on broad development theory, and the Millenium Development Goals. Palo et al examine trends outside the forests sector, and the importance of global corporations involved with forests. They also review via discussants issues such as climate change, forests and water, and globalization. From these investigations they conclude that it would be unwise to attempt to derive a specific set of policy prescriptions capable of dealing with all these issues, but also an optimal combination of market instruments and policies will need to be considered and implemented in specific cases; in other words, these things cannot be simply left to chance, or delegated completely to agencies or authorities outside the sector.

As we will argue later in this book, what happens to forests is very much defined in a broader context of economic and social change, and the way that this context is evolving is itself subject to change. Obviously, the count of studies and publications devoted to this subject is an order of magnitude greater than that of material related directly to forests, and we can clearly do no justice to this body of work in this book, although we reference elements of it throughout. The current United Nations targets for development for 2015 are embodied in the Millenium Development Goals but, as a cursory reading of the general literature and analysis of this subject will reveal, there are many shades of opinion as to what these goals really mean, and how they should be implemented. A sense of this debate can be obtained from a recent book by Andrew Sumner and Meera Tiwari (2009). They point out that advocates of the MDG approach see it as a multi-dimensional approach to poverty reduction, rather than measuring success simply in terms of aggregate GDP growth. Sceptics, however, maintain that the approach has not evolved from being a donor-led, reductionist agenda that ignores local ownership of the process. Greater emphasis, they suggest, needs to be given to what the poor actually say is important to them; security, respect, voice and similar concerns can be more important than consumption, in some cases.

1.2.2 There Are Still Some Things Left to Say

Despite all that has been said in the various fora and publications on global forestry, we believe that some of the basic questions surrounding what needs to be *done* have often been subsumed under a tide of partial and uncoordinated solutions, or assumed away under ideological judgements and determinations. The avalanche of available written material has, if anything, made the task of making sense of forests issues – or even in some cases deciding what these issues are – more difficult and confusing. Because opinions and even basic information on the subject vary widely according to source, there is a need to move some distance back from the subject when attempting to develop an overview and some perspective on it, so that the reader can assess what is of vital importance, and what less so.

This approach will require some selectivity about the issues and the information used; there is no other way to retain a broad perspective on the world's forests. There will occasionally be some deconstruction of myths and legends, aimed at revealing new possibilities and approaches, or in some cases resurrecting old ones. The authors' positions (and, no doubt, prejudices) have been forged by several decades each of experience in the sector: To some extent, their manifestation may reflect the varying states of mind – ranging from amused, through amazed, and annoyed, to (occasionally) aghast – which tend to manifest in those on their journey through these issues.

As will already be apparent, this book begins in a rather unusual way, by introducing first, as a core issue, the nature of the debate about natural forests in developing countries which has raged for decades now. In Chapter 2, we will elaborate upon our view that the fragmented and dysfunctional nature of this debate, centred around the basic forests sustainability issue, has led to a vicious cycle of events: lack of significant success in the field with sustaining forests, to competing assignments of blame among the various interest groups and stakeholders involved at the international and national levels, leading back to alternative approaches and solutions, and thence (so far, at least) to another round of failures in the field.

Large questions about the purpose and nature of forests sustainability underlie the debate about what to do in the forests: should we want to sustain all or most of the large forests left in the developing countries? What do we mean by sustainability, in the forests context? How should it be done? Who should do it? Who should pay for it? Who should benefit? These are the questions (in very reduced form) which have been at the centre of much of the controversy and debate which surrounds forests in the developing world.

1.3 The Dynamics of Forest Loss

Before we can proceed with diagnosis and remedies for the forest loss problem in this book, we need to have an appreciation of the scale and directions of the forest loss problem. It is for this reason that Part II of this book, comprising Chapters 3–5, addresses:

- The state of the world's forests: an analysis which goes beyond the obvious global decline in natural forests, to explore the regional variations and some of the global implications of continued forest loss.
- The role of trade demand for forest products in determining the status and condition of natural forests. It will be argued that there can be both negative and positive effects on forest cover resulting from trade, and that this fact is another source of controversy around the forests sectors of developing countries.
- The impact of other activities, often grouped under the term *drivers of deforestation*, on natural forest cover.

1.3.1 Natural Forests Are in Decline

The world's natural forests are certainly in decline[1]: The Food and Agriculture Organization of the United Nations (FAO 2006) estimates a gross global loss of forest areas between 1990 and 2000 of 13.1 million hectares per year, and 12.9 million hectares per year between 2001 and 2005.

As will be discussed in Chapter 3, these are approximate estimates: it is easier to estimate net changes in forest area (i.e. allowing for plantations (on both initially naturally forested land, and open sites) regeneration of natural forest and so on, but for our purposes in this book, plantation and pre-existing forests need to be treated very differently: factors such as carbon content, biodiversity, and some other broad natural resource implications are important in this differentiation, as are basic wood volume changes (plantations grow much more quickly than most natural forests: a hectare of plantation can easily produce ten or even twenty times more wood over time than occurs on a hectare of natural forest).

Our broad objective in examining this question will not be to present another analytical projection of what is going to happen on forest sustainability, but rather to summarize what seems to be the current wisdom on this, and to draw some conclusions about what this means for our subject.

---

[1] As will be seen in Chapter 3, there are various technical arguments about what constitutes a forest, as opposed to more open woodland, and debates about where forest degradation ends, and actual deforestation begins, but in any event wherever this definitional line is drawn, the statement on decline still holds.

As is usual in almost any area to do with natural forests, there is controversy and disagreement present in the issue of global forest loss: Other estimates of forest loss examined in Chapter 3 derived by different means will provide alternative approaches to the measurement and interpretation of the loss of forests.

1.3.2 So, Is the World Running Out of Wood?

In the 1980s and to some extent the 1990s, the main question surrounding forest loss was whether the world would soon be running out of wood: from an international perspective, wood, for use in manufacture of solid wood products and pulp and paper was seen as an important question, especially in view of the declining state of the world's natural forests resources.

More recently, it has been observed that booming consumption in large emerging economies such as China and India, increased illegal logging, over-harvesting in some key supply zones and a number of other relatively recent developments may be making the earlier forecasts of relatively low consumption (and therefore price) pressure on global wood supplies less convincing.

On the other hand, it has been suggested that many developing countries will reduce their dependence on fuelwood and charcoal as a source of energy (presently these uses consume about as much wood volume as total industrial wood), thus leaving room for more production from natural forests available for industrial use. However, it is questionable how significant this factor will be, given that most fuelwood used comes from parts of trees not used for wood products manufacture, or from trees grown in areas where commercial manufacture of wood products is not viable.

The focus has since shifted to the ecosystem value of natural forests, as plantations seem to rapidly be becoming more important as a source of roundwood for wood and paper products. Strong arguments have been put forward that there are rapidly increasing demands for other, non-wood goods and services that forests provide and that specific regions will in fact encounter shortages. Forests can produce multiple goods and services, but there are trade-offs between how much of each can come from a given area of forest.

When considering forest loss, there is also the question of forest degradation, which can occur in forests due to over-logging or other inappropriate forms of use, and which leads to deforestation, but which is poorly measured and in some cases poorly understood. This issue will be raised in the context of Reduced Emissions from Deforestation and forest Degradation (REDD), in Chapter 7 of this book.

There are also important regional and national differences in forest loss or degradation. The great majority of deforestation is occurring in developing countries; most developed countries are not actively reducing their forest cover – in some cases if only because they have more or less completed the effective removal of their forest resources a century or more ago – and indeed most are now engaged in increasing forest cover within their borders.

The Food and Agriculture Organization of the United Nations has argued that in developing countries the combination of persistent widespread poverty, high population growth, low productivity and poor capital formation in agriculture, and the potentially adverse impacts of global warming on agricultural development over the next 50 years, will continue to place great pressure on natural forest areas as a source of new agricultural land areas[2]. As will be discussed in Chapter 6 of this book, the poverty issue has been revisited as a causal factor, and there are more complex underlying political and institutional reasons for excessive deforestation. A re-focusing of the priorities in how governments and development agencies seek to address this is urgently called for.

1.3.3 Can Developing International Trade and Natural Forests Survival Be Compatible?

In very general terms, at the present time, there are very few forecasts which actually predict steeply rising real prices for forest products on global markets as a result of wood shortages

In Chapter 4 of this book, the role of the international market for forest products is examined, and conclusions about the effectiveness of this market to call forth supplies, and to develop both internal and external substitution alternatives for wood products are drawn. The chapter addresses the broad question of whether large natural forests can survive in the face of international trade in forest products. We consider developments and views on current forest sector trade and its impacts on forests, and we will retain this disaggregation of the trade subject in the following chapters of this book.

We will consider whether, if and when serious global shortages of industrial wood do begin to develop, markets will respond adequately, and whether in the medium term there is likely to be a discernible impact on global economic activity and growth from such developments.

Trade is also changing in nature: production of many forest products has been shifting to developing countries, including those which have large forest resources. Sensitivities as to the sustainability of wood from developing countries – especially those with tropical rainforests – have stimulated demand for certification. Certification is a market-based, non-regulatory forest conservation tool designed to recognize and promote environmentally-responsible forestry and sustainability of forest resources. The certification process involves an evaluation of management planning and forestry practices by a third-party according to an agreed-upon set of standards. It appears to be in large measure a response to perceived corruption and inefficiency in the sector; a lack of confidence in some governments as custodians of forests, on the part of some interest groups and segments of the public – especially in consumer countries.

---

[2] There is some debate on the FAO figures: The Science and Development website cites some satellite imaging covering the 1990–1997 period which result in figures for annual deforestation 23% lower than those provided by FAO.

There is no reliable information on how widespread this perception actually is but, judging on how eagerly state forest agencies from the United States to Russia, from Sweden to Malaysia, have embarked on independent certification of their forest management, the concern certainly seem to be prevalent.

Certification standards address social and economic welfare as well as environmental protection. of wood products in some developed country markets, but there have been, and remain, serious technical and political difficulties associated with the certification process. Certification was conceived as a two-pronged initiative, to promote both sustainable consumption patterns and sustainable forest management, but as will be discussed in this book, it has been beset by unresolved disagreements between rival certification systems, and it also has not produced market premiums for certified products at the scale originally conceived. At this point, therefore, the overall impact on forests of certification is still marginal and for many managers of natural tropical forests achieving certification is an uphill battle, involvement significant costs.

The chapter also considers the impacts of trade liberalization on the state of forests, and on the economies of forest rich developing countries. These are serious questions for both developed and developing countries to consider, in the overall context of preventing excessive deforestation. The issue of the legality of harvested logs is a vexed one in many forest-rich countries, but, while it may seem counter-intuitive, there may be some danger that a preoccupation with this could in fact damage the attainment of sustainability.

1.3.4 What Else Is Happening to Natural Forests?

The larger problem of natural forest loss is created, in a sense, by the range and ubiquity of goods and services that these forests provide. Multiple use management has been a standard catchphrase in the management of forests in developed and developing countries for many years now, but it is easier said than done. Many possible uses of natural forests are completely incompatible with each other (for example, complete protection of biodiversity and other ecological properties of a natural forest, and logging – even sustainable and selective logging) and most are interlinked in some way, so that an increase in the intensity of one form of use will exercise downward effects on the availability of others.

In Chapter 7 we provide a brief list of the various non-wood products and ecological services that rainforests produce; many of them of fundamental livelihood significance to the hundreds of millions of people who live in or near these forests. The medicinal value of many rainforest plants has been well-known to communities living in or near these forests for centuries, but it is also a fact that some 25% of modern pharmaceuticals are based on compounds extracted from rainforest plants, and significant elements of the gene stocks which have produced fast-growing crops and other plants originated from rainforests. Of course, there are also many non-wood products and forest goods and services that can be derived from forest types other than rainforest.

Virtually any area of natural forest that is relatively accessible and stocked with goods and/or ecological services of some commercial or livelihood value will be under competition for use, often by groups of stakeholders who neither agree with nor in many cases approve of each other. These uses range from fuelwood and non-timber forest products[3] used primarily by local communities, through local and regional environmental services provided by forests, such as water catchments and soil stabilization activities, to their so-called global forest values, as repositories of valuable biodiversity and gene banks, and stored carbon – a crucial element in climate change mitigation. There are major concerns for the sustainability of all these, and this will be an important subject for discussion in this book.

It is the interplay of these multiple possibilities for what might be done with forests or, as we will see, with the land upon which they are located, at levels of usage which are in aggregate unsustainable, which leads to the set of drivers of deforestation, both direct, and indirect. Discussion of drivers needs to consider not just the activities themselves, and their impacts on the forest, but also the incentives involved, and we will spend some time on this later in this book. Moreover, we need to deal with the fact that the nature of deforestation itself in many developing countries has also evolved, and continues to do so. In the past, it was largely associated with poverty and local subsistence activities such as shift-and-burn agriculture and fuelwood collection.

The situation has changed, however, in countries such as Brazil and Indonesia, which have become the most important locations as far as deforestation is concerned. In these countries, the movement of commercial scale agriculture and grazing onto previously forested lands has become a significant phenomenon in recent years. This suggests a development that has more direct implications for developed countries than previous forms of deforestation, for it is in developed country markets where the bulk of products produced in this way are consumed, and so the question of what measures to address this particular problem needs to be taken by those consumer countries arises.

An important caveat to this idea will, however, be raised in this book: assignment of an overwhelming causality of forest loss to these cash crops is potentially misleading: it cannot be assumed that what finally ends up on a given area of previously forested land – whether it be oil palm, soy bean crops or any other of the many possibilities – is necessarily the only, or even the predominant cause of that particular deforestation. We will discuss these issues in Chapter 5 of this book, and will review some field data to assess the realities of deforestation, and to speculate on what measures and, importantly, what financial resources might be required to address the problem comprehensively.

---

[3] Medicinal plants, mushrooms, berries, other fruits, etc.

1.4 Setting the Scene for Sustainability: Valuation; and Financing of the Forests

The consideration of the dynamics of forest loss in Part II of this book, as outlined above, will help provide a basis for assessing the potential forest cover situation. However, even when the measurement and definitional issues outlined above are dealt with, there are more fundamental questions which need to be addressed: primarily, what level of reduction in deforestation – globally as well as for each country specifically – is possible? How could those be achieved? Would such an effort be worth it? Who will want to pay?

Part III of this book will address these questions. Chapter 6 will open the discussion by addressing some general issues of environmental sustainability, historically and in today's world, and will then move on to the sustainability of the remaining great natural forests. Many will argue, of course, that the answer to the question of how much deforestation should be accepted should be zero: the world has already lost too much valuable natural forest. Obviously this is not an answer: it is an assertion. Others will assert that further losses are inevitable, for reasons such as those outlined by FAO, and that the primary concern must be to ensure that converting forests to other forms of land use must be economically and socially worthwhile, and environmentally responsible (which has certainly not always been the case in the past).

In this part of the book, we will be suggesting that while this latter approach would represent common sense under the assumption that the global boundary conditions for investment in forests are unlikely to change, it is precisely these conditions that need to change, and there are now significant prospects that this could come about.

A critical observation made in this part of our book is that while forests might be an obvious case of natural resource sustainability being sacrificed to the cause of economic growth, it is by no means the only one. We provide a brief review of the history of debate about environmental sustainability, beginning with Malthus, and ending with a snapshot of recent events suggesting that the limits to that growth – prematurely and incorrectly invoked during the nineteenth and twentieth centuries – is now beginning to look more real, especially if we do what economists sometimes like to do, which is to assume everything surrounding the growth dynamic itself remains constant. This will, of course, sound to some like the story of the shepherd boy who cried *wolf!* many times, alarming the local villagers for his own amusement. But of course the denouement of that particular fable is that when the boy finally did sight a wolf heading for the local flock, his cries of warning were ignored.

This is not to suggest that there has been any shortage of international and intergovernmental consultations, agreements, treaties on the broad question of environmental sustainability, and in Chapter 6 we review the history of these often large and well-publicized events. This process is by no means over, and while many disappointments and failures can be identified in it, it is reasonable to suggest that a focus on specific solutions – rather than the broad canvas of radical and rather theoretical reform – has begun to assert itself through the process, and the associated question of realistic options for financing change has been increasingly considered.

This is all to the good, because if it is the case, as observed earlier, that the world may be approaching the point where environmental degradation becomes a serious threat to continued growth and prosperity – i.e. the wolf finally being in sight – then we will have arrived at the point where a new paradigm of growth, one that will require environmental and natural resources sustainability to be treated as real factors in economic growth projections, instead of being placed in that category of difficult realities or "constraints" routinely paid lip service, and then studiously ignored.

1.5 What Has Happened to Forests Sustainability So Far?

The glib answer to this question, for the forested countries we are concerned with in this book, would be – not much. However, this conceals the fact that a great deal of effort, and considerable innovation, has been invested in the effort on sustainable forests. We need to look for answers that offer more in the way of potential solutions than this one. As the reader may have already guessed, this will lead us directly back into the dysfunctionality and conflict inherent in the international forests constituency, for if there is one subject that will reliably re-open the door to these phenomena, it is the subject of forest sustainability.

We demonstrate the nature of this dilemma in this book by way of a real-world example of how a highly ideological approach to forest sustainability created a sustained and highly unproductive conflict within the international forests constituency; one that has repercussions still resounding through the corridors of the World Bank, and other major development agencies.

Chapter 6 closes with reference to some of the important lessons which have been learned, in the long and in many ways disappointing interaction between the governments and other internal stakeholders around forests in the countries which have large forests remaining on the one hand, and the development agencies, developed country governments, environmental and social NGOs and other interest groups that from part of the international forests constituency on the other.

1.5.1 Valuing Forests: The Full Ecosystem; and the Carbon Values

One of those critical lessons which has been learned is that, for as long as the perception of the value of a standing natural forest (compared to the alternative uses under rapid exploitation of the marketable forest products followed by conversion of the forest to other uses) remains very different between the various stakeholders and interest groups involved, the default condition will remain continued deforestation. This book is based on the idea that large scale forest loss and degradation in developing countries cannot usefully be discussed in isolation from valuation of forests as a whole, nor from macro-level economic and non forest sectoral policies which have an impact on forests, nor from much larger global issues to do with global warming abatement and international resource security issues.

High levels of international and in some forested countries, national concern about the uncontrolled loss of rainforests in particular has led many analysts and forest conservation advocates to attempt calculations of the economic value of forest ecosystems, and we will review some of these efforts in Chapter 7 of the book. Reference to the earlier discussion about the multiple use of forests, and to the brief listing of goods and services that intact forests can provide will serve to illustrate the immense difficulty of such quantification, and unsurprisingly, therefore, it has proven to be a controversial area of debate. The review of this area of analysis will show that the calculations of total ecosystem value tend to develop enormous estimates of the worth of these forests. Even in cases where these estimates have been successfully defended against methodological criticisms, the reality is that at the present time, no market exists which could monetize these values at the scale calculated.

However, recent developments and rising concern about global warming have re-invigorated the discussions of valuation of forests, but this time on the basis that the immense stores of carbon sequestered in the biomass of intact natural forests (and in this area, the tropical rainforests are the predominating provider) *is* potentially a marketable commodity, in a world where large emitters of greenhouse gases may soon be required to either reduce their emissions directly, or make acceptable investments in carbon-positive activities as offsets to the emissions. As is now well-known, carbon emissions from forest loss globally are believed to be in the order of 15–20% of all human-induced emissions. We will cite some recent data and analysis in this book which will show that it will be extremely difficult for the world as a whole to reduce emissions sufficiently to slow and eventually reverse concentrations of atmospheric carbon levels to the extent believed to be necessary to avoid serious, possibly unmanageable events as a result of warming, unless loss of natural forest areas – especially the tropical rainforests – is reduced very significantly. It will be equally necessary then that those forests are managed sustainably with special attention to the carbon outcomes of that, and, where possible, forest areas are increased as well.

We will also review some recent global estimates of the likely costs of reducing deforestation, in comparison to the potential size and value of the carbon emissions reduction market, and we will add some considerations of our own into this mix, with a view to drawing conclusions as to the economic and political feasibility of a global scale avoided deforestation carbon offset market developing in the near future. We will briefly review the issues involved in implementing and measuring avoided or reduced deforestation and degradation, and we will discuss the options that will need to be considered in future intergovernmental meetings on this issue.

1.5.2 Financing Change in the Forests

The final topic to be considered in our book, prior to drawing some conclusions and recommendations from our considerations of the large issues involved in global forests, is that of financing the change to reduced loss of important forest values and better management in the forests of the developing countries. In a very real sense, this issue underlies most if not all of the broad observations we have made in this book about this loss. We will defend our opinion that, regardless of whatever else will be needed to bring about significant reductions in forest loss, very significant increases in financing will be required to drive this change. The failure of efforts of the international donor community, and governments of the countries with large forests themselves to bring about improvements in sustainable forest management and conservation in developing country forests to date is essentially a failure in appropriate valuation of the suite of benefits forests provide, which in turn has led to ineffective financing efforts.

We will review four basic options (not mutually exclusive) for financing major change in the natural forests:

- Improving the scale and effectiveness of the historical development agency donor government partnership based approach
- Integrating forest issues more effectively into the macroeconomic planning and budgeting process in countries with high endowments of natural forest
- Implementing an effective international forest carbon trading market, and the monitoring and verification measures needed to validate this, under the regulatory framework for Reduced Emissions from Deforestation and forest Degradation (REDD) to be considered under the United Nations Framework Convention on Climate Change (UNFCCC) processes
- Initiating early adoption of reduced deforestation and protection of the full suite of forest ecosystem goods and services, through a combined public-private sector financing of the necessary activities, driven by underlying rather than immediately marketable carbon values.

References

Humphreys, D. (2006) *Forest Politics: The Evolution of International Cooperation*. James and James-Earthscan
Humphreys D (1996) Forest politics. Earthscan, London
FAO (2006). Global forest resource assessment *2005*, FAO Forestry Paper 147. Rome
Moelino M, Wollenberg E, Limberg G (eds) (2008) The decentralization of forest governance: Politics, economics and the fight for control of forests in Indonesian Borneo. Earthscan, London
Sayer J, Maginnis S (eds) (2007) Forests in landscapes: Ecosystem approaches to sustainability. Earthscan, London
Sumner A, Tiwari M (2009) The millennium development goals and beyond: New directions in poverty reduction and development policy. Palgrave Macmillan, Basingstoke
Palo M, Uusivuori J, Mery G (2001) World forests, markets and policies. Kluwer, The Netherlands
Vanclay JK, Prabhu R, Sinclair F (2006) Realising community futures: a practical guide to harnessing natural resources. Earthscan, Sterling, London, VA World Bank. (2004) Sustaining Forests: A Development Strategy. The World Bank, Washington, DC
World Bank (2004) Sustaining Forests: A Devolopment Strategy. The World Bank, Washington, DC

Chapter 2
Global Forests: Debate and Dysfunction

Abstract This chapter addresses the dysfunctionality of the international dialogue on natural forests – especially those in the tropics. It argues that, until very recently, the issue of global deforestation has lost considerable traction with the broader public, and that disputes between different groups within the international forests constituency – usually over ideological issues – have reduced the impact of development assistance financing on forest management and protection, and in fact have reduced the willingness of some development agencies to engage with the forests issue at all. The danger is that if this fragmentation of the approach to reducing deforestation is not addressed, then the same fate may be in store for promising forestry options from climate change mitigation that will be discussed later in the book. The essential task must be to persuade *all* groups which have significant agency over forest outcomes – not just those groups of most direct interest to the particular organization promoting its own solutions and programmes – that their natural forests really can be made worth more dead than alive.

Forests cover 30–40% of the total land area of the earth (the figure varies depending on the definition of forests used; FAO[1] estimates global forest area at 3,869 million hectares in year 2000: about 30%), and contain about two thirds of the leaf area of all land plants, and about 70% of the carbon contained in all living things (Table 3.1 in the next chapter summarizes some basic global forest information at regional level). They contribute directly to the livelihoods of around a quarter of the people living on this planet.

---

[1] By far the largest repository of global forest resource information is held by the United Nations Food and Agriculture Organization (FAO) which periodically publishes global forest resource assessments: the latest of these covers the period 1990–2005. These estimates will be discussed further in Chapter 3.

There are three large – and interrelated – propositions which we want to present and defend in this book:

- The sustainability of natural forests in the developing world is of global – not merely national – environmental and economic importance, but to date has not been treated as such by the international community.
- A fundamental divide exists between the perceptions of value of standing forest held by those international, national and local interest groups who have had most agency in terms of maintaining current or historical patterns of use of these forests, and those held by groups who have sought to bring change in forest outcomes primarily through advocacy, among which we include major international donor agencies and multilateral development banks, for financial and other reasons we will elaborate upon.
- The achievements of the advocacy approach – which probably never had promising prospects as a primary instrument of change in the forests – have been further weakened by internal conflict, agenda-splitting and fragmentation in the international efforts that have been applied. A new paradigm is needed, based on financing sustainability at a scale concomitant with the rate of natural forest loss and in a manner which engenders consensus rather than its opposite.

We believe we are at a crucial tipping-point, in the future of the world's forests. On the one hand, there is now a strong prospect that the new paradigm referred to above could develop, due to the advent of the forest carbon issue as climate change rises in public concern globally. However, the same problem of internal conflict within the international forests constituency which has minimized progress on sustainable management of forest to date, will threaten the benefits that could come from effectively addressing forest loss, forest carbon and broader forest ecosystem concerns, unless awareness of what is at stake, and effective negotiation in place of set-piece political conflicts are developed.

To make our case for these propositions, as noted in Chapter 1 above, we will need to cover a very wide range of issues in this book, from the status and condition of the forests themselves, through the role of international trade and aid in forest outcomes, across to new and intriguing developments in global carbon markets and their potential impacts on forests (and vice versa), with many other technical and political way stations to be visited on this journey.

We begin, in this chapter, by laying out in more detail our interpretation and definition of the major issues and questions to be addressed. The chapter closes with an outline of how we will structure the book to address these in more depth in the later chapters.

2.1 Defining the Problem #1: A Dysfunctional Dialogue

As will by now be apparent, a major motivating factor for the authors in preparing this book has been a sense that something is badly wrong in the forests of the developing world, that too much natural forest area is being lost or degraded, and that, so far, not enough is being done about this. The authors are aware that they share this feeling with many others in what might loosely be termed the international forests constituency: this book will avoid using the more common term "community" for this group because, as will become apparent, it is anything but communal in its functioning. It is in fact a disparate web of government officials in forested countries (and, for that matter, in countries whose forests are depleted or naturally poor and inadequate for the country's needs), local stakeholders including forest communities and smallholders, people from development assistance agencies, multilateral development banks, environmental and social non-governmental organizations, forest industries, academics, researchers, and a large number of other interest groups. It would be fair to say that – with some exceptions – the mood on forests in this community is predominantly one of unease and uncertainty, ranging through in some cases to profound gloom.

Despite the concern manifested within this global constituency, until very recently the forests issue has probably lost traction as a matter of global importance with the broader public in recent years, and this is a critical problem. Forests have had to compete with other global concerns: the international economic crisis; continuing concerns with terrorism and security; and the potential macro-consequences of global warming itself.

The latter is somewhat ironic because, as will be examined in greater depth in Chapter 7 of this book, global warming should be driving a new focus on the primary importance of retaining forests. Until quite recently, however, the complexity and in some respects the intractability of the process of developing an effective global approach to greenhouse gas (GHG) abatement has delayed focus what is often termed the low-hanging fruit of field responses to the abatement task, of which we believe reducing deforestation in natural forests is one.

This might become a particularly potent constraint in the case of forests, because as noted in the second of our broad underlying ideas in this book as outlined at the beginning of this chapter, people and interest groups involved in natural forests have different values and perceptions on the role of forests in society, and there are strong conflicts of opinion present. There is no common view on: what, or who, is causing the problem; what can and should be done about this; who should do it; and who should pay the bill. These basic questions have grown into a jungle of issues and controversies which range from basic technical arguments as how forests should be managed for conservation and sustainable use, through ideological and philosophical differences over the role and importance of forests, relative to everything else, to major institutional confrontations. In the latter case, it is obvious that an organization established to finance and manage implementation of forestry projects will have a very different outlook to one which exists primarily to raise public awareness as to the perilous state of the natural forests and of those people most dependent upon them for basic livelihood.

Dialogue must somehow operate effectively across all the levels noted above, if significant change is to occur, but our concern is that so far, the dialogue has become hostage to the fraught political environment and the entrenched positions protagonists in it have adopted. Forests, especially as seen through the various prisms of the international forests constituency, have become a hunting ground for narrow and ideological special interest groups, and for contrarians of all stripes.

As this free-for-all has unfolded, many onlookers who are interested and concerned about forests have become almost afraid to ask: *what do we need to do?* This is because in certain circles of the constituency, it has become virtually politically incorrect to pose this basic question – Many of the interest groups involved proceed from the basis that they already know what needs to be done, and the only relevant question for them is how their solutions can most comprehensively be applied. A major objective of this book is to find a way through the thickets of issues and arguments that abound, back to the basic question of what needs to be done.

The fundamental task is to arrive at practical solutions, and to focus the international policy dialogue effectively on these. We will therefore attempt to engage the reader in a dialogue on what is possible and what may be desirable. In this process, we will need at times to take a broad broom to the detritus of symbols, metaphors, codes, political tropes, silver bullet solutions, single-issue fetishes and shibboleths which have accreted, piece by piece, like irregular bricks into a growing wall of incomprehension that surrounds the forest sector.

2.1.1 The Problem You See Depends on Where You Stand

This calls to mind the old aphorism that, to a person with a hammer, everything looks like a nail. What various groups and agencies interested in natural forests see as major issues, problems and viable solutions varies widely. We can best illustrate this by briefly outlining the perceptions of the major interest groups involved:

- *Ecology first.* This group views the current situation in forests from a fairly holistic ecological perspective, and (often) concludes that the loss of large areas of natural forests which is occurring is simply unacceptable to the world community on any terms.
- *Global public goods matter.* This group suggests that biodiversity and stored carbon (which, if released, will add significantly to greenhouse gas emissions) – are of much higher value to the world at large than is currently reflected in the way these forests are used and protected.
- *Social uplift from forests.* This group is mainly interested in social progress and poverty alleviation, advocating the primary importance of forests for local communities, especially as an effective means to generate income and employment for and alleviate poverty in these communities.
- *Indigenous forest dwellers.* This group attempts to define and then defend the rights and interests of indigenous people heavily dependent on their traditional forest habitats, as being of paramount concern in areas where this is an issue.
- *Forests are for multiple use.* This group centres on professional managers of state-owned forests and advocates the multiple use of forests, combining commercial scale forest logging operations with allowances for indigenous and community uses, and a range of environmental purposes such as watershed and biodiversity protection, all under a broad, state-mandated regulatory system.

Indirectly, this group can include urban dwellers, who consume significant amounts of various of the goods and services forests may provide.
- *Private forest owners.* This group is concerned about protecting their constitutional rights in the management of their forest assets for their own objectives, involving maintaining financial profitability under regulatory requirements for sustainability and forest protection.
- *Loggers and processors.* Private logging and processing sector interest groups may exhibit more sympathy for the private forest owners and state forest managers than do most of the other interest groups outlined in this list, since they tend to be the vehicle through which private access to the commercial harvesting of wood in forests is provided.
- *Macro-economic and social policy planners and decision-makers.* This group comprises economic and social policy decision makers at both local and national government levels. Inclusion of this group in a listing of forest interest groups requires some explanation, since it will be apparent that many people operating in this group would fail what would appear to be a basic criterion: that of having any strong interest in forest outcomes. However, while they may have relatively little concern for or knowledge of forests per se, their decisions and actions can have profound effects on these natural resources, and on people dependent upon them, and this group must be engaged effectively by forests interest groups before real change in outcomes is possible. The corollary to this is that a significant proportion of the consequences of broad economic and social policy decisions in many economic systems ends up being manifested in the natural resources sector, whether or not this is known by the perpetrators of such decisions, or by the forest interest groups involved in the country.

This brief listing of the interest groups involved in forests should suffice to demonstrate the complexity and political sensitivity involved in this issue, and the potential for conflict among these groups. As it has turned out, the plethora of views and interests abroad in the sector has led to loss of focus and fragmentation in international and national views and policies, and this has been a major constraint on the effectiveness of the international forests constituency and its many dialogues on forests to come up with real and workable solutions to what most see as undesirable forest loss. All too often, opportunities for consensus on fundamental issues have been lost in a tempest of debates on tiny details, as the various interest groups promote their causes and, often, re-define reality to comply with their views.

2.1.2 The Search for Perspective: "You Talk; We Chop"

This is not to suggest that solutions will be easy to find: it does suggest, however, that a rational perspective must be sought. As has been noted, the economic, social and environmental issues that underlie the objective of sustaining forests *are* enormously complex, and almost irredeemably interwoven into broad issues of national political and social reform, economic development, and even larger ones of global stability and prosperity. Because of the intensity of the differences in views held, much of what has been (and is being) written and said by various interest groups on sustaining forests tends to place the forests issue (or whatever aspect of it is being discussed) at the centre of a large political and policy universe, and has then focused rather vaguely on what *someone, somewhere* – national or local governments, communities, the private sector, environmental advocacy groups, the global community – should want (or be compelled) to do about forests, and then on how they should go about it.

When this constraining forest-centric lens is applied, one predictable casualty is a cogent analysis of exactly how relevant decision makers should be identified, and then persuaded that (a) the issues identified really are important at national or international scale, and (b) that the recommended course of action will succeed. As a result, some of the initial material written on this subject can appear as the triumph of optimism over experience. Later in the process, it can end up wanly concluding that there is an apparent lack of political will on the part of those with the power to do something about the problems identified – an observation which seems to occur as something of an afterthought, coming as a rather unpleasant surprise to the proponents, rather than having been, in many cases, entirely predictable.

What is often not recognized, in this context, is that the *framing* of an argument is often as important to gaining perspective, as its detailed substance. In time, many of the current debates within the international forests constituency on what is an appropriate outcome in a given case, or what sector based approaches will best achieve this, will almost inevitably appear quite secondary in the larger scheme of things.

Consider, for example, the case of deforestation. It is apparent that most interest groups operating at national or international level probably agree that too much natural forest is being lost – or, at least, that much of what is being lost is going out at too low a price. However, there will be some influential opinion operating in an economy at some remove from the forests sector which will suggest that losing something in the order of a quarter or even a half of one percent of a given national forest estate annually[2] is unfortunate, perhaps, but of relatively little moment in the larger task of economic development. Unless alternative arguments can be put directly to those holding this view that the costs of ignoring or under-valuing forest outcomes, and the benefits of an alternative sustainable approach, will both be much higher than may seem presently to be the case, then this view will prevail.

Unfortunately, however, many of those involved in the sector who could be formulating those arguments often seem to spend more time and energy arguing with each other as to what *is* the most effective means of reducing deforestation, and who should be liable for what in implementing this. This then sets off a

---

[2] See the deforestation figures cited under the discussion of global forest resources earlier, and at more length in Chapter 3.

multiplicity of arguments, which have not at this point been convincingly resolved. Similarly, as we will see in Chapter 5 of this book, the subject of the causes of deforestation remains a fraught and political one, where ideology (i.e. what is the more convenient cause, in terms of making the argument for remedies a given group is interested in pursuing) is often in the ascendancy over ideas (i.e. what is really happening in the forests in a given area, and what will work best to address the *real* causes – as opposed to symptoms of deforestation, which are often misidentified as causes).

In recent years, as will be discussed later in this chapter and in more detail in Chapter 7 of this book, large global public goods considerations, such as the value of forests as repositories for biodiversity, and carbon sinks (a highly relevant subject in view of the climate change issue) have entered the discussion – but the emphasis has often been on identifying these large benefits, rather than on the much more important questions of how their production will be financed, and how the benefits from doing so will be distributed. This subject is crucial to the issue of survival of global natural forests.

An illustration of the point being made here – that discussion and debate need to gain both focus and perspective in the forests arena, if things are to change for the better, comes from a response made by a logging company manager operating in South East Asia, when declining an invitation to join the energetic and seemingly endless debates of the international forests community as to what should happen to the forests of his region: *"You talk; we chop."*

2.1.3 Re-evaluation, or Late Onset Apostasy?

Some readers will view many of the messages in this book as late-onset apostasy on the part of the authors. To some extent this might be so, although we would hope that those who know us better will recognize some of our dissents as items of long standing in our lexicon. In other words, our commitment to the conventional faith and wisdom in some critical areas in the forests arena has been in doubt for some considerable time.

Apostasy, in any event, has a long and creditable history in this vexed subject of global forests. One famous example of it is drawn upon as inspiration for this present work:

In 1962, Jack Westoby, already then well-known as a forest economic thinker, put down what was by then the conventional approach to forests and forestry, based on the prevailing theory of economic development which suggested that poor nations should emulate rich nations, by accelerating their industrialization and putting their natural resource capital to work in pursuit of this objective as quickly as possible. Forests are a good way to do this, because of their strong forward and backward linkages to other parts of the economy, acting as growth multipliers.

Later in his career, at a time when most eminent men are content to rest on their laurels, Westoby did something quite unusual, although typical of him: He recanted on his views on forests and development, launching a fierce and broad-ranging attack on his own previous position. In a ringing denunciation at the World Forestry Congress in Indonesia in 1978 (published in the proceedings of that event, and in Westoby 1985), he berated the developed world and argued that poor countries are underdeveloped as a consequence of the development of rich nations: the success of the latter group is founded on failure of the former, and is sustained by it. In case any who heard him should take away any false notion that he intended to exempt forest industries from this swingeing attack, he made it clear that, in his view:

> ....very, very few of the forest industries which have been established in the underdeveloped countries have made any contribution whatever to raising the welfare of the urban and rural masses, have in any way promoted socio-economic development. The fundamental reason is that those industries have been set up to earn a certain rate of profit, not to satisfy a range of basic, popular needs.

Westoby's turnaround was a bombshell. This was a prominent man; one who had written and thought long about this sector and its relationship with the rest of the world. It doesn't really matter whether one favours his earlier view or the one he replaced it with, or if, in the world of the early twenty-first century, merits and flaws can be perceived in both. What Westoby achieved at the time was a quantum leap in intellectual questioning of the basic beliefs and habits of thought which had directed forest development policy.

Westoby was not a natural cynic – he was basically a humanitarian, and humanitarians generally make poor cynics. His doubts were forced on him by what he saw, and in the process he made people involved at many levels in the sector think. In recent decades, many people have attempted to claim paternity of the idea that real people – not just anonymous corporate or public sector functionaries – must be regarded as the primary beneficiaries of what is done in and with forests, and that this must be something which can be sustained as part of the overall development process. As was often the case, Westoby got there first, and this idea for forests is his legacy.

Through his own intellectual status, and the drama of his repudiation of previous verities, he was able to attract wide public interest in the debate. As noted earlier, the quality of this debate – especially as it has reached the public – has waned in recent years, as fragmentation of opinion amongst interest groups has proliferated.

2.1.4 Re-focusing the Discussion

Without any illusions as to matching the impact of Westoby's Damascene experience, an aim of this book is to stimulate some basic re-thinking within the international forests constituency, and hopefully to bring about more effective re-engagement between that constituency and the broader public on forests, forest dependent people, and what is to happen to them as a result, since, ultimately, the participation of that broader public will be the driving force for change. This will involve a re-focusing of discussion on the realities of forest use, loss and protection in today's world, and an examination of some of the options available for dealing effectively with the most pressing problems. There is presently some considerable distance to go to reach this state of affairs.

To achieve the quantum leap needed to close the perception gap, all involved must be prepared not just to speculate on what in aggregate we all lose if large natural forests diminish or disappear, but also on what it will really cost us to *avoid* this happening, and on who will be expected to pay these costs. In the real world, there will be *opportunity costs*[3] associated with almost any course of action. Because uses and outcomes in forests are so bound to events elsewhere in the national economy (and international trade), a decision to manage a piece of forest in perpetuity as a natural forest may not be justified just because the direct benefits of doing so exceed the direct costs, axiomatic though this may seem to some operating within the sector. Keeping the forest intact usually implies forgoing the opportunity to use the land there for alternative purposes, each of which will have its own costs, benefits and indirect flow-on effects on people, the economy and the environment associated with it.

Unless there exists a fair idea of what these may be for the most obvious alternatives, the opportunity costs of the alternative chosen are simply not fully understood. A common outcome on naturally forested land is unsustainable forest exploitation practices, often leading to degradation of the resource. Unsavoury as it will seem to some, this outcome is an alternative to sustainable forestry, and should be evaluated as such along with all other alternatives. In practice, this has rarely occurred.

This becomes even more complicated when the problem is generalized to the multitude of real-world forest types and situations that characterize the world's developing countries: the enormous disparities in practices, endowments, social and economic conditions and governance and political realities that confront them. If we then add in the disparate objectives and priorities of local, national and international interest groups involved in the sector we will soon find that, as acknowledged earlier, making substantial progress in saving the global forests – if indeed that is our aim – will not be simple.

2.2 Defining the Problem #2: Sustainability and Forests Value – The Basic Issues

It is necessary, when seeking an understanding of how sustainability plays out in the forests, to address some fundamental questions related to the history and development of the concepts of economic and environmental sustainability as a whole. In Chapter 6 of this book we will review some of the history and issues involved in

---

[3] **Opportunity cost** is the *cost* of taking some action or purchasing something in terms of the benefits of the *opportunity foregone* – in other words, of other investments or actions that could have been taken instead. The concept of opportunity cost is one of the key differences between *economic cost* and *accounting cost*. It is fundamental to assessing the *true cost* of any course of action, because it brings out the hidden costs of an economic decision.

this subject: concerns of natural resource limitations first raised by Malthus, and again in the 1970s by the Club of Rome, and Paul Ehrlich, as well as repudiations of these (principally by economists).

2.2.1 What Have Forests To Do with Global Economics?

To any who might be questioning the connection between this large subject and the forests issue, two snapshots of how views on the subject of resource sustainability have emerged will illustrate this point. Both, co-incidentally, come from people who have occupied the position of Chief Economist in the World Bank, and then moved onto larger things:

Consider the following quotation:

> There are no ... limits to the carrying capacity of the Earth that are likely to bind at any time in the foreseeable future. There isn't a risk of an apocalypse due to global warming or anything else. The idea that the world is headed over an abyss is profoundly wrong. The idea that we should put limits on growth because of some natural limit is a profound error and one that, were it ever to prove influential, would have staggering social costs.

Were one to come across a statement such as this with a relatively recent date on it – say any time after the year 2000 – the informed reader would be inclined to ascribe it to some troglodytic market fundamentalist blogger, or to the (dwindling) group of large energy companies that feel the need to propound this view[4]. These operations play to an audience of journalists and politicians of the general persuasion that the whole climate change issue is a sinister fabrication of groups wishing to de-rail the inspiring march of capitalism.

In fact, this statement – widely quoted and cited ever since in publications and websites – was made in the early 1990s by Lawrence Summers, a highly respected economist who was Chief Economist for the World Bank at the time, and who went on to become US Secretary of the Treasury, and then President of Harvard University, and now chairs the National Economic Council for the Obama administration in Washington. The purpose of citing this statement here is less to remind ourselves of the considerable risks of acts of intellectual hubris on large and fast-moving subjects, than to point up the degree to which mainstream concern about global environmental stability has developed in the intervening period. It would be highly unlikely that Summers himself would express support for these views at the present time.

---

[4]Usually from behind the veil of manufactured non-government organizations or think tanks that have been created for this purpose. See, for example, the Competitive Enterprise Institute, and the Heartland Institute, which appear on websites in the guise of a think-tanks, promoting the views of climate sceptics and various free market fundamentalist views. CEI even solicits public donations for it to carry on its chosen tasks, but in fact it was founded on generous funding from Exxon-Mobil, and other large corporations.

Consider next the conclusions of the landmark report on the economics of climate change for the British Treasury produced in 2006 by a team led by Nicholas Stern – now Lord Stern – who earlier was also Chief Economist for the World Bank. Stern's report (Stern 2007) as is now well-known, is a salutary warning that the global costs of continued high levels of greenhouse gas emissions from human activity will be extremely high by mid century, if nothing is done to reduce these emissions in the meantime, and will continue to grow beyond that point. Stern makes the point that the loss of natural forests contributes more to this human-induced emissions growth, and thereby global warming, than all of the emissions from the global transport sector.

This will serve to demonstrate the increasing recognition and growing public awareness of the strong connection between natural resource sustainability and prospects for continued economic growth, and the important part that natural forests play in this.

2.2.2 A Question of Value

One of the reasons why halting or slowing deforestation in many of the natural forests of the developing world has proven so difficult is that assigning reasonable value – monetary or otherwise – to all the goods and services that forests can provide is problematical, and controversial. However, any valuation made is of little consequence if all interest groups expected to be involved in the sustainable use and protection of forests do not actually *receive* some value that they regard as significant and satisfactory, from doing so.

As will become evident in this book, the *scale* of financial flows into tropical forests that would be needed to convince national governments, local governments and local communities that it would be truly worthwhile for them to achieve sustainable management is huge: it will amount to billions of dollars of additional funding each year. Nothing close to this level of funding has to date been available as grants or other concessionary financing from developed nations. Nor has it come from the fruits of rapid economic development based on forest management in the developing countries, or at least, not until considerable time has elapsed, and much – probably most – of the forest resource has been lost.

2.2.3 Sustainability: The Elusive Objective

Sustainability has become the watchword of the debate over natural resources in recent years, especially as serious water constraints, extreme weather events, large-scale forest fires and other major developments are associated ever more closely with climate change in the public mind. In forests, the sustainability concept has been bandied about for longer, and in the last two decades or so, a fierce debate has raged on the subject among environmental groups, national governments of richly forested countries, forest owners and managers, forest industries, local communities and their advocates, international development agencies, and many other interested participants. The following extract from a paper written in 1999 by the late Alf Leslie (Leslie 1999) – one of international forestry's great iconoclasts – will illustrate the combative mood which has developed:

> There are two unavoidable facts of life that have to be faced in the whole issue of the conservation of the tropical forests, which has now subsumed the development issue. The first, despite all the hypocrisy with which the world tries to ignore it, is that conservation depends, above all, on providing a guaranteed and permanent livelihood to the hundreds of millions of people who, in their present conditions, have no choice but to keep clearing forests to grow food. The second, despite all the propaganda to the contrary, is that forest-based industrialization is one of the few means of doing this on the scale and with the continuity needed. In effect, like it or not, sustained yield management and utilization of the tropical forests for industrial wood is a necessary condition for their conservation. This is not to say that forest-based industrialization can deal with all of the poverty that underlies deforestation; it can make a sizeable contribution but it is not the whole solution.

Historically, as Leslie suggested, in the area of forest sustainability hypocrisy has certainly been in evidence in some of the efforts of forest-owning governments, large donors and development banks, environmental and social NGOs, the private sector and other interest groups. So, while there may now be a possibility of a way out of the dilemma Leslie identified 10 years ago, as the forest carbon issue emerges, caution and realism need to be continually deployed.

2.2.4 *Why Has Sustainable Forest Management Not Worked?*

As the Aristotlean debates over the true meaning of sustainable forest management and how it should be achieved have moved ponderously along in the international forests constituency, almost any acceptable version of it in the field in developing countries (especially those in the tropics) has proven very hard to come by. In fact many of the same issues on the sustainability question are awaiting clear answers in developed countries.

These observations are not intended to denigrate the genuine efforts being made in the many interesting, innovative and positive experiments on sustainability taking place, by many people who are passionate, dedicated and committed. The problem arises because the integration, prioritization and therefore the general implementation of the good work being done at pilot scale is not happening on a large scale, for reasons of lack of tangible incentive. The marginalisation of forests as an issue on the global stage will continue for as long as this remains the case.

Sustainable forest management has succeeded in many places (usually those with adequate resources and options to allow this to happen), but has manifestly failed in many others. The failure of sustainable management of natural tropical forests in particular has been attributed to problems of unbalanced vested interest and related inadequacies of sector governance, including corruption, the poor performance of public forest agencies and the private sector in many countries.

2.2 Defining the Problem #2: Sustainability and Forests Value – The Basic Issues

While these shortcomings certainly are present in many cases, we suggest they are symptoms, rather than the underlying cause of the failure of sustainable forest management: Sustainable management of natural forests has failed in many parts of the world not because the technologies of managing forests in this way are non-viable, nor because the goals of management systems in place do not recognize or understand or support all elements of sustainability. It has failed because of two central issues of policy and incentive in the political economy of this sector:

- First, many of the economic and social policies influencing forests and forest dependent people are initiated a long way from the forest sector itself and can only effectively be manipulated by mechanisms that operate well outside the sector. Government policies designed to promote the development of exports based on a large non-renewable natural resource (oil is the classic example), or to intensify agriculture, or develop domestic manufacturing behind tariff and non-tariff protection policies, are cases in point. Development policies have often been adopted for reasons that have little or nothing to do with concerns about forests, but which can have major impacts (which can be good or bad) on forests. Another case is where changes in external trade conditions for a given country creates an incentive for rapid expansion of agriculture[5]. This can also have good or bad impacts on forests, although historically it seems to have been primarily the latter.

    It is likely that the dynamics of such large economic and social changes have been determined at senior levels of economic policymaking, in response to the political agenda of the government, or to forces originating outside the country: Either way, the reality in many countries is that these changes have almost certainly not involved prior consultation with the interest groups most concerned with forest outcomes – in many cases not even with the ministers for forests or for environment themselves. Bringing in these groups after the fact, in an attempt to mitigate adverse impacts on forests resulting from such large scale exogenous changes, is likely to be an ineffective – or at least an inefficient – approach. Changing perceptions on forest values held by the individuals and groups who are actually responsible for national economic and social policies would be more successful, but this would require very different approaches on the part of forest sector interest groups.

- Second, there is a group of incentive issues related to important stakeholders who *are* closely involved in the forests – government forest and environmental agencies, forest owners, private sector operators, local communities, and others – any (or all) of whom have not been convinced of the broad and long term benefits of sustaining forests, or who cannot afford it due to lack of immediate livelihood alternatives. These groups will not have made the necessary compromises to their own interests, nor will they have been sufficiently driven by public concern over the consequences of not pursuing it to do so.

---

[5] Rapid expansion of soya growing in Brazil, or oil palm plantation in Southeast Asia, are examples, and we examine these (among others) in Chapter 6 of this book.

It is especially important to acknowledge, in this context, that agencies and institutions responsible for forests at the national and local government level in many cases have good reasons not to be convinced of the economic and social benefits of sustainable forest management.

Values for the *global public goods* that can come from forests – biodiversity, and increasingly now the value of carbon sequestered in the biomass of natural forests – have often been the forests issues which are uppermost in the minds of international groups contemplating the consequences of forest loss. However, these concerns have not been included in the economic analyses of the agencies and actors involved in decision-making about forests at national and local level.

Unfortunately, until now, there have been very few practical opportunities available to access the *large scale* international financial transfers that would reasonably compensate governments and local populations for the supply (via retaining forests) of the global public goods that these forests can provide, and at this juncture this lack has to be seen as a failure of the international forests constituency to deliver effective solutions to excessive forest loss.

The same reasoning applies when communities are given ownership and/or management of forests, although in some of the literature on sustainability of forests, this form of tenure and management is often claimed to be a critical reform that in itself will produce greater sustainability; a useful compilation of the issues and arguments involved in this subject can be found in (White and Martin 2002). It is undoubtedly true that in many cases where large scale industrial operations occur under state forest management, with primacy of access accorded to commercial harvesting operations, these can seriously disadvantage local communities, economically and socially. This creates perverse incentives for these communities to encroach upon forests in a destructive and unsustainable way, since no other avenue is available to them to benefit from the forests. In effect, the forest in this situation takes on the role of a de facto commons, with the usual fate of an open access resource, i.e. degradation.

This much is commonly noted and discussed in the dialogues on community forestry. What is less recognized, however, is that redressing this situation by transferring the forests to community ownership, whatever this may achieve for their social and economic well-being, will not *necessarily* provide any more forest sustainability than would other forms of tenure. Regardless of ownership, specific incentives to encourage such a long term approach, instead of lucrative and unsustainable economic exploitation, will need to be in place for those involved, to achieve sustainable forest management, and we will take this issue up in Chapters 6–8 of this book.

None of the above arguments is intended to suggest that the inclusion of *other* values embodied in forests is not important in decision-making about them. The inclusion of reasonable values for catchment protection, locally consumed forest goods and services and the many other benefits provided by forests which do not enter formal markets (and therefore only rarely enter the calculations of economists and policy makers), can be of real importance to governments and local communities, and can and should therefore be added to the case for sustaining forests from their viewpoint. However, later in this book we will argue that in the absence of large scale transfers to suppliers for provision of global public goods from these forests, the task of changing the mindsets of those currently involved in the use, management and conservation of forests, those (such as local communities) who might in future assume a more prominent role in these same activities, and those responsible for much larger economic and social decisions which can impact upon them, becomes radically more difficult.

2.2.5 Never Mind the Differences on Sustainability: Can't We All Just Get Along?

In the circumstances outlined above, the approach adopted in this book to determining priorities for the international forests community is to recommend examination of what changes will actually be required, inside and outside the sector itself, and at national and international scale, for *significant* improvements to forest management to be made. From this, some judgement can be made as to whether, and where, such change will be worth pursuing, from the economic, environmental and social viewpoints.

The objective is to show how development assistance agencies, environmental groups, local interest and civil society representatives, forest owners and managers and progressive-minded industry might best co-operate on significant improvements in specific situations: *but* at significant scale. We will return to the matter of financing this scale of operations in Chapter 7 of this book, and to the necessity to frame this approach in terms of basic social change in Chapter 8.

2.2.6 Asking the Wrong Questions

In the absence of such an approach, it has remained tempting (for some) to conclude that a thousand blossoms should bloom on the forests sustainability question – that all approaches to retaining as much natural forest as possible have merit, and should be tried. This may seem intuitively correct, but we argue that the need for selectivity and prioritisation in this area is critical, and trumps the broad spectrum approach. Grounds for this argument could be established by examining the merits of all possibilities here but at this juncture, it will be more effective to consider two seemingly simple questions, to illustrate the case:

"How much biodiversity is enough?" This question is not apocryphal: it was actually asked of one of the authors of this book, by a senior Planning Ministry official in Indonesia, during an earnest discussion of what the priorities for forests in Indonesia should be, and what organizations like the World Bank, major donor agencies, NGOs and other interest groups should be doing to contribute most effectively to achieving them. It is a question which certainly needs to be answered practically in all countries of the world.

To some people involved in this field, and many looking on, the answer would seem simple: *all of what is left must be protected.* This is definitive, certainly, but will sound to the harried official more like ideology than a practical answer. Others may respond with: *as much as can be achieved*, as being the appropriate answer. But this is not an answer, it is simply a new question: how much *can* realistically be achieved?

The fact is that for any country, even the broadest question of how much of a given natural resource (let alone one specific element or dimension of it, such as biodiversity is of forests) is enough, is an exceptionally difficult one to answer. Natural resources are a source of capital – the universally scarce resource in poorer countries seeking economic development. We have already foreshadowed the point that unsustainable use of natural resources in pursuit of economic growth is nothing new in economic history: what is new is that the global limitations on this strategy are becoming rapidly more evident. Even in situations where serious problems of governance and fiscal management are not present in a country, using such resources to generate investments in the economy too quickly can be hazardous, from the viewpoint of the stability of the environment or the social system which ultimately underpins the sustainability of economic progress.

On the other hand, using the resources *too* slowly can impede the prospects of economic development and alleviating poverty, raising another set of stability issues in the form of social and political risks that no government will willingly engender.

This line of reasoning underscores one of the major interests in this book on the subject of global forests, and what is to happen to them. Correctly defining a target for biodiversity, or any set of goods and services provided by forests, is difficult under the best of circumstances. It involves assessing all the needs related to natural resource use (employing appropriate criteria and indicators and the best available scientific and technical information); valuing or prioritising them (including opportunity costs) to the extent possible; consulting with stakeholders; and then taking firm decisions. Attempting to set targets before this is done, and the long list of relevant factors are understood and built into the response, is akin to writing a shopping list in complete ignorance of both the cost of the items included on it, and the funds available to make the purchases.

How then can we best help to achieve sustainable forests management? This question is commonly put, and often then acted upon, by donors concerned about forest loss, but it suffers from the same problem of prior assumptions being made as is the case in the biodiversity question examined above: The question implicitly assumes sustaining forests is accepted by almost all interest groups in a given country and situation as an unreservedly good thing, and that assistance or intervention in the forest should therefore take this position as a starting point. However, as we have seen, this will only be actually true when the divergence between national and international interests on the real value forests themselves is reduced significantly. If all reasonable values and benefits flowing from forests to all interest groups are taken into account, then sustaining these forests into perpetuity would be seen by all as the optimal approach in many cases (although the same process of prioritization and valuation outlined above for biodiversity would need to be applied in each case here as well before this conclusion is drawn).

However, this is a long way from being true at present. The problem, as noted earlier in this chapter, is that in practice functioning markets and governance systems which would allow developing country governments and local communities to capture and distribute a reasonable share of the total benefits from sustainable management and protection of forests are not often present. Until they are, significant levels of forest degradation or deforestation will be tolerated. This is nothing new: the ongoing exploitation of tropical forests at the present time is repeating a development-driven transformation of the landscape historically experienced by developed countries with significant forests. Only the pace is different.

2.2.7 The Right Question

There is another question we could ask that would usefully replace those we have posed on biodiversity and sustainable forest management above. It is: *What will need to change for people and groups who are most influential in determining outcomes in forests to believe that sustainably managing and protecting larger areas of forests is a worthwhile objective?* Answering this question would require that consideration be given to broadening the group of interlocutors who would need to be involved, beyond the direct interest groups in the forests sector, and that the nature of the approaches commonly used in pursuit of forests sustainability be reconsidered.

2.2.8 Implications for Some Popular Current Approaches to Sustainability

If the primary question becomes how adequate incentives for SFM to those who will actually make decisions affecting forests, can be provided, then it can be argued that the approaches to forest sustainability that are prevalent in some of the donor and financing agencies in the international forests constituency will not achieve this.

This is because there is a logical disconnect between what many in the international forests constituency identify as a crucial element in achieving sustainable forests management – the political will within the country to do so (and most observers will agree strongly that this is necessary) – and what they go on to argue will *produce* this: improved governance in the forest sector. However, improving governance in the sector *per se* will not necessarily lead directly to the manifestation of political will needed further up the line to bring significant areas of forest under sustainable forest management and effective protection. There is no convincing evidence of this causality in the field at present. Improving governance may (probably will) assist governments or other groups achieving their real objectives, but this does not mean that it will change what these objectives themselves are, and as suggested earlier, they may well be ones which are not at all consistent with those held by some other members of the international forests constituency.

Therefore many of the approaches to forests sustainability related to governance that are presently supported by international forests interest groups from many developed countries – attempts to reduce or eliminate corruption, decentralizing the management and ownership of forests to lower levels of government and/or local communities, introducing measures to eliminate or reduce illegal logging and excess forest processing capacity – are unlikely to succeed in the longer term in significantly reducing the amounts of forest being lost in many countries, unless political commitment to this goal at the highest level of government is clear and unequivocal.

This does not mean that these things are not in themselves desirable goals to pursue: there is little doubt that better forest sector governance, reduced corruption, effective decentralization and empowerment of local communities in forests, and a rational approach to illegal logging, would in most cases have beneficial impacts on overall economic efficiency, social equity and poverty alleviation, and we make the case for these in many places throughout this book. But, undertaken in isolation, or in limited form – as they often are – they will lead to little progress with the larger problem of achieving sustainable management of forests. So, while they may each be considered a *necessary* condition for improving sustainable management and protection of forests, none will amount to a *sufficient* condition: The point here is that they may in themselves do little or nothing to reduce forest loss – either for areas which could be managed under sustainable forest management, or areas that should be completely protected for their high conservation and environmental values – *until* the fundamental questions of linking large economic and social changes and reforms to forest outcomes, and reconciling incentives among different stakeholders in forests, are resolved.

2.3 The Global Dialogue on Forests: Moribunds, Mercantilists, and Manicheans

Much of what has been discussed so far has dealt with the perceptions held by various groups or agents within the international forests community on what is wrong in the forests, and what needs to be done. It will probably be clear by now that the argument being advanced here is that the discourse on forests is beset with problems of irrelevance, lack of focus and prioritisation, and fragmentation into a large number of specific issues, often isolated or at best only tenuously related to other developments going on in the sector. The result is that the whole – in terms of a cohesive view on what matters and what does not, and, importantly, what is most necessary to do – has become much less than the sum of its parts. The dialogues and communications within the international forests community and, even more importantly, between it and the wider public, have become dysfunctional.

Perhaps the most obvious disappointment, in this regard, involves the moribund nature of various attempts to marshal international support at government level for agreement and codification of measures to support management and protection of forests that have been made over the last two or three decades, and the history of these will be briefly reviewed in Chapter 6.

Forests are located in the territories of various countries, and are of great and immediate economic value to their governments. They provide livelihood to large numbers of people living in or near them. Attempts by international bodies or interest groups to impose or even energetically negotiate controls over the way these forests are exploited have been met with strong resistance – and will continue to do so, until real progress is made in reducing the gap in different stakeholders' perceived value of forests, discussed earlier in this chapter. The concept of forests as providers of global public goods remains, at this stage, an aspirational one: one that would have more possibility of success if the focus were put on developing the means to compensate individual countries and communities for the provision of the global public goods which these forests can provide, rather than on attempts to draft unwilling and sceptical governments into unfunded mandates and broad agreements, if such benefits are not part of the package.

The response of many forest-rich developing countries to the attempted moral pressures on them arising from the global dialogue on forests, has been to resort to a potent form of mercantilism, whereby central regulation of the process of use of the forest was introduced to legitimize a domestic agenda based, in reality, on heavy and unsustainable exploitation of the forests. This tendency is exhibited in developed and developing countries alike. David Suzuki, the well-known environmentalist broadcaster and author has shown signs of frustration at the lack of progress in alerting the world to the dangers he sees for forests. In a newspaper article in 2005,[6] he was reported as noting that resource extracting communities seem to regard natural resources as merely an economic opportunity, and show little interest in issues of sustainability.

Had the international dialogue operated at a level of sophistication and pragmatism that was needed to do justice to the controversial and emotive subject of forests, then things may have turned out very differently. In fact, it has focused on the more trenchantly partisan (and newsworthy) exchanges on forests, and has therefore not reflected the quite significant level of moderation and reasoned internal debate that actually takes place in and around the sector. In general the mass media which, by definition, has to simplify its messages, has been successfully attracted to associating wood harvesting (of any kind) with forest destruction. A particularly appealing subject has been indigenous people who are suffering from loss of habitat due to expanding frontier of forest use (together with many other encroaching activities such as mining, tourism, and so on).

---

[6] *Thunder Bay Chronicle*, June 4, 2005.

The term Manichean has often been applied – incorrectly – to describing a relatively straightforward conflict situation, with a simple choice between good and evil being on offer. Much of the international dialogue on forests has had this nature. It is an irony that the original Manicheans were in fact orthodox dualists who believed that the forces of Good and Evil were *not* engaged in a large messianic struggle, but rather that their contrasting presence was the very basis of the spiritual order. For the Manicheans, this dualism constituted the structure of the spiritual world that framed each individual's relationship with reality: people should examine the Evil within themselves, and take a personal journey towards Good dominating Evil, but recognising that evil can never be eradicated. The international dialogue on forests might have made far better progress had this real Manicheanism been more in evidence.

In reality, as we have suggested, while the relentlessly adversarial nature of the public dialogue generated considerable public interest at the time, it has in the longer run contributed to a decline in both the quality and reach of the public debate on this issue. The reason why people in Canada may not be getting Suzuki's message is certainly *not* a lack of information, analysis and opinion on issues in forests: it is that there are so *many* competing messages out there on this issue, and they are often vested interest-driven, cacophonous, contradictory and, from the public perspective, confusing. Nothing is more likely to erode public interest than this combination.

It is (or should be) a matter of serious concern that the dialogue on forests, and what should happen to them, has for some time been quite unconstructive – or in some cases just plain toxic. This may not have been the intention of all groups who have been drawn into this process, but it has certainly been the result. It has happened at the very time when a consensus among those involved in some way and the interested general public, on what should be done, needs to become much stronger, and more united.

The issues involved in this subject are enormous in scale, and the audience that needs to be involved in considering them equally so, if the necessary momentum for changing the situation and incentives in the sector is to be generated.

As already noted (and elaborated upon in Chapters 5–7 of this book) there are major changes developing in the international market for forest ecosystem services, due largely to one of them – sequestered carbon – being a central element in the wider issue of climate change and greenhouse gas emission reduction. Dealt with correctly, this could lead to a new paradigm for international forestry, especially for the large rainforest owning countries. It is to be fervently hoped that the same fragmentation and conflict characteristics which have dogged the effectiveness of international forests dialogue to date will not persist in this new situation.

We are of the view that the means by which this hope can be realized is to engage a broader audience than that which has been defined earlier as the international forests constituency, and to bring this wider group more effectively into the discussion and debate over what now needs to be done. It will only be when this is achieved that political support for real and constructive change can occur. For reasons we will examine in this book, there are new dynamics related to climate change and increasing public recognition of the imperatives of global environmental stability which could open opportunities for much greater engagement in the global forests issues we have introduced in this first part of our book.

References

Leslie AJ (1999) in Reflections, Unasylva 182, Food and Agriculture Organization of the United Nations, 1999

Stern NH (2007) The economics of climate change: The stern report. Cambridge Press, Cambridge

Westoby J (1985) Foresters and politics. Paper presented to the Ninth World Forestry Congress, published in Commonw Forestry Rev 64(2):105–116

White A, Martin A (2002) Who owns the world's forests? Forest tenure and forests in transition. Forest Trends and Center for Environmental Law, Washington, DC

Chapter 3
The State of Global Forest Resources*

Abstract This chapter presents data and observations from the literature on recent trends is forest cover globally, and the current state of the resource. It briefly canvasses some of the issues and debates related to defining forest, and to the means by which estimates of forest cover have been, and should be, prepared – a processes which reveals some fundamental differences in approach on this question.

The particular case of tropical rainforests is raised specifically, mainly to draw attention to some recent findings on the very high ecosystem value of these forests at a global scale; findings which emphasize the potential seriousness of the continued high rates of loss of tropical rainforests globally.

The chapter concludes by making the case that even though attempts to value natural forest ecosystems in the recent past have failed to convince, it is also arguable that the conventional economic wisdom which holds that natural forest sustainability is generally not an economic proposition is also wrong. Proper recognition of the real risks associated with large scale forest loss, and the implementation of more imaginative means to bring some of the ecosystem values into effective markets, are essential when examining the sustainability question.

As we saw in Chapter 1, the issue of how much of forest biodiversity the world needs, and how we should go about protecting what remains, is a vexed question, and we will be returning to aspects of it throughout this book. It will come as no surprise that the broader questions of how much forest overall is needed, what role humanity has played in the reduction of forests so far, and what we should do about this now, are no less difficult to address, complicated as they are by major concerns of sovereignty, trade, and global environmental stability – the latter being an issue which has gained considerable momentum as concern about global warming has developed.

---

*The authors wish to acknowledge the assistance of Professor Rodney Keenan, Head of the Department of Forest and Ecosystem Science, University of Melbourne, Victoria, and Director of the Victorian Centre for Climate Change Adaptation Research Centre, for his assistance and advice with preparation of this book, and especially this chapter. Our views on the subject as expressed in this chapter are not necessarily those of Professor Keenan.

We will return, later in this book, to the issue of the carbon value in natural forests, in the context of all other forest goods and services. As will become apparent then, our belief is that the carbon value of natural forests is the one that has the potential to be a key game-changer in the struggle to maintain and protect the surviving global natural forests; the other qualities and benefits of those forests will, in a sense, be joint products with the carbon good. For now, we ask the reader to bear with us on this matter.

The subjects raised in the above paragraph alone would easily take up more space than the entirety of this book, and we cannot do justice to all of them here. We will instead draw upon published data and analysis to address two specific questions which are of central importance to the arguments and observations we have on global forests in the remainder of this book:

The first is: *what is the state of the global forests?* Obviously, this question has been, and remains, a major subject in international forestry and the debate over the environmental aspects of forest loss. To answer it, we need information on what the base store of natural forests is at the present time, and what rate of decline (or expansion) is present in that resource, and we will need to build some historical context around this issue.

Here, we need to invoke yet another caveat on our coverage: The question of the overall state of the world's forests resolves through two issues – the rate at which naturally occurring forests are being reduced and/or regenerated; and the rate at which plantation forests are being established, and utilized. As we will elaborate upon later in this chapter, our interest is much more in the natural forest area, and less in the plantation one. This is by no means to suggest that the plantation subject is not important and interesting. It is most certainly both, but it involves a very different set of issues – many of them more closely related to agricultural development, than to the management and stewardship of the naturally occurring forests. Even the instrument under which creation of new forests through afforestation and reforestation are addressed under the United Nations Framework Convention on Climate Change (UNFCCC) – the Clean Development Mechanism – is quite different to what is envisaged under the REDD (reducing emissions from deforestation and forest degradation in developing countries) mechanism, which is presently under consideration under the UNFCCC deliberations, and about which we will have a good deal more to say in later chapters of this book. The two approaches may well be dealt with under a single forest instrument under the UNFCCC eventually, but they will nevertheless be quite different activities.

The second broad question that we are especially interested in here is: *what will be the implications of addressing deforestation for the global approach to environmental sustainability and economic development?* Although this question has, as we will see throughout this book, generated a great deal of study, analysis, debate – and controversy – we will suggest that the forest carbon question, and its role in the approach(es) taken to ameliorating global warming, has cast a new light on this subject, and we need to be abreast of this development.

3.1 The State of the World's Forests

3.1.1 The Historical Picture

It is not easy to assess the impact of humanity on forests over the long term. There have been attempts to measure the state of forests in the pre-human era, and then to estimate the impacts of subsequent human activity in forests, but this has proven to be a complex and difficult problem. One limitation has been the dearth of precise data on exactly what condition the forests were in at various points in the past – at least until the advent of satellite imagery, which has allowed very large scale assessments of resources to be made, and changes in that assessment through time to be estimated. Williams (2008) has noted that even into relatively modern times, forest clearing was regarded as unexceptional, and little was written or recorded about it. He notes that historical estimates of pre-human forest cover range from 2.4 to 6.4 billion hectares globally, with a convergence around 4.1 billion hectares. He estimates that the total loss of forest cover over human history due to human activities is in the order of 700–900 million hectares, or about 19–22% of the original total. This figure is significantly less than some estimates of relatively recent losses in specific areas – the Amazon, the Indonesian forests and so on – might suggest.

Defining forest. One of the major problems which arises when discussing global forest resources is how forest is actually defined. Numerous versions of this are currently in use. Definitions vary widely from country to country and between agencies within countries. These can generally be grouped into three categories (Lund 2002): those describing administrative units, those that describe land cover and those that indicate a type of land use. Administrative definitions are often linked to the land base rather than the vegetation cover and therefore land may be defined as forest but have no tree cover. Definitions based on land cover generally define threshold values: minimum area, strip width, canopy cover, and tree height. Definitions linked to land use sometimes use a minimum threshold of production as an indicator of the capacity of the land to provide timber or some other commodity or sometimes simply specify that the land must be under 'forest use' (for example, the Food and Agriculture Organization of the United Nations (FAO) definition, which is primarily designed to exclude agricultural tree crops).

FAO has been seeking to establish an international consensus on such definitional questions since it first began to collect information from member countries for the purpose of its periodic Global Forest Resource Assessments in 1947. The following definition was used for the 2005 Assessment (FAO 2006):

> Land spanning more than 0.5 hectares with trees higher than 5 meters and a canopy cover of more than 10 percent, or trees able to reach these thresholds in situ. It does not include land that is predominantly under agricultural or urban land use.

This definition provides for a high degree of inclusion: countries with only small, fragmented areas of tree cover with low stocking or crown cover can still report some area of forest. However, it is regarded by many as too low, particularly in terms of the minimum crown cover, which is generally below the minimum level that most ecologists would consider the minimum threshold where trees constitute the dominant life form (Williams 2008) and it is a difficult threshold to detect using satellite-based remotely-sensed data.

3.1.2 Global Forest Cover and Cover Change

FAO has been issuing assessments of global forests since 1980: The Global Forest Resources Assessment (GFRA) 2005 (FAO 2006)[1], is the most comprehensive of these to date – not only in terms of the number of countries and people involved, but also in terms of scope. It examines the current status and recent trends for about 40 variables covering the extent, condition, uses and values of forests and other wooded land, with the aim of assessing all benefits from forest resources. Information has been collated from 229 countries and territories for three points in time: 1990, 2000 and 2005. The results are presented according to six thematic elements of sustainable forest management.

Total forest area as of 2005 according to the FAO definition was estimated at 3.95 billion hectares or 30% of global land area (0.62 ha per person). 'Other wooded land' (generally land with trees but not meeting the minimum definitional thresholds) was 1.38 billion hectares (FAO 2006). This gives a total forest and wooded land area similar to previous estimates of the total area of denser forest (Williams op cit).

For the 2005 GFRA, countries provided information on their forest area for three points in time. The net change calculated through this process is therefore the sum of losses due to deforestation or natural disasters and increases in forest area due to afforestation and natural expansion of forests. The total net loss in forest area over the 10 years from 1990 to 2000 was −8.9 million hectares per year. This rate of net loss reduced somewhat in the period between 2000 and 2005 to −7.3 million hectares per year. The FAO did not have capacity to disaggregate these estimates into absolute values of loss and gain in each country. An estimate of total loss was generated by summing the loss for countries with a negative change in forest area. This resulted in a gross loss of forest of 13.1 million hectares per year between 1990 and 2000 and 12.9 million hectares per year between 2000 and 2005.

The FAO figures are very approximate estimates of gross forest loss, which cannot be monitored effectively, while the net change in forest area is easily detectable based on remote sensing data. The actual rate of deforestation will be higher than indicated in the above figures, as these do not include the forest area lost in those countries that reported a net increase in forest cover or the areas that are balanced by regrowth in those that reported a decrease.

---

[1]The summary of results presented here draws heavily on this document from FAO (2006).

Forests are not evenly distributed across countries. The Russian Federation alone accounts for 20% of the total area and five countries (the Russian Federation, Brazil, Canada, the United States and China) account for more than half of total global forest area (2,097 million hectares). Twenty two countries account for about 80% of the total forest area. Seven countries or regions (mostly small island states and other nations with small territories) had no areas that qualified as forests using the FAO definition and there are 64 'low forest cover' countries with less than 10% of their total land area covered by forests.

At a regional level, South America has the highest percentage of forest cover, followed by Europe and North and Central America. Asia has the lowest percentage of forest cover.

Table 3.1 utilizes FAO resources data to summarize at regional level forest and woodland area, and deforestation.

Forest loss rates are not evenly distributed: The largest net losses were in South America, about 4.3 million hectares per year between 2000 and 2005, followed by Africa, where countries collectively reported a net loss of 4.0 million hectares per year. Net loss was about 0.35 million hectares per year in total in North and Central America and Oceania. There was a decrease in the rate of net loss in Oceania, with a reduction in the rate of loss in the primary contributor, Australia. Forest loss increased in North and Central America, primarily due to a lower plantation establishment rate in the United States (down from 596,900 ha per year in 1990–2000 to 157,400 ha per year in the period 2000–2005) and a slower net rate of forest loss in Mexico.

Table 3.2 shows estimates of forest carbon and production figures from the world's forests.

Ten countries with the largest net loss per year in the period 2000–2005 had a combined net loss of forest area of 8.2 million hectares per year:

The ten countries with the largest net gain of forest area per year in the period 2000–2005 had a combined net gain of 5.1 million hectares per year due to afforestation and reforestation efforts and natural expansion of forests. China has recorded large increases in forest area, due to recent large-scale afforestation programmes for soil and water protection and combating desertification as well as increasing production of wood and non-wood forest products.

Predictably, the subject of measuring forest cover and forest loss has been controversial in the literature: there are many definitional and methodological issues upon which technical opinion differs widely. In some earlier exchanges, Achard et al. (2002) argued on the basis of utilizing global imagery that forest loss in the humid tropics between the years 1990 and 1997 were 23% lower than generally accepted estimates of deforestation for this area. Fearnside and Laurance (2003) took issue with a number of assumptions and omissions by Achard et al., suggesting that the cumulative effect of these led to a large underestimate of deforestation in this zone.

More recently, Grainger (2008) has argued that claims in the FAO studies that tropical forests are declining is not supported by the evidence. Grainger has examined the UN data, and claims to have discovered many errors and inconsistencies in

Table 3.1 Forest and woodland area and deforestation by regions (FAO 2006 Global Forest Resources Assessment)

Region/Area	Forest area 2005 1,000 ha	Other wooded land 2005 1,000 ha	Total forest and woodland 1,000 ha	Annual deforestation 2000–2005 1,000	Deforestation rate 2000–2005 %/year
Total Eastern and Southern Africa	226,534	167,023	393,557	−1,702	−15.3
Total Northern Africa	131,049	94,609	225,658	−981	−3.5
Total Western and Central Africa	277,829	144,469	422,298	−1,356	−16
Total Africa	635,412	406,101	1,041,513	−4,039	−34.8
Total East Asia	244,862	90,003	334,865	3,839	−0.6
Total South and South-east Asia	772,850	209,848	982,698	4,828	−15.3
Total Western and Central Asia	43,587	71,446	115,033	13	6
Total Asia	1,061,299	371,297	1,432,596	8,680	−9.9
Total Europe	1,001,393	100,924	1,102,317	661	18.8
Total Caribbean	5,974	1,311	7,285	55	−0.8
Total Central America	22,411	5,018	27,429	−285	−7.4
Total North America	677,464	111,867	789,331	−101	−0.3
Total North and Central America	699,875	116,885	816,760	−386	−7.7
Total Oceania	206,255	429,908	636,163	−355	−5.6
Total South America	831,540	129,410	960,950	−4,253	−3.2
Total World	4,441,748	1,555,836	5,997,584	363	−43.2

3.1 The State of the World's Forests

Table 3.2 Forest carbon, production forest area, removals by region (FAO 2006)

Region/Area	Forest carbon 2005 Million tons C	Production forest area 1,000 ha	Industrial roundwood removals 2005 1,000 m³ o.b.	Fuelwood removals 2005 1,000 m³ o.b.	Total removals 2005 1,000 m³ o.b.
Total Eastern and Southern Africa	16,067	41,067.309	34,185	151,242	185,427
Total Northern Africa	3,908	44,172.497	8,138	173,007	181,145
Total Western and Central Africa	53,038	52,788.917	36,303	266,919	303,222
Total Africa	73,013	138,028.72	78,626	591,168	669,794
Total East Asia	10,147	125,547.99	114,658	56,326	170,984
Total South and South-east Asia	53,592	371,225.31	273,543	225,540	499,083
Total Western and Central Asia	2,172	9,668.322	14,667	19,500	34,167
Total Asia	65,911	506,441.62	402,868	301366	704,234
Total Europe	206,162	723,915.46	542,882	138,513	681,395
Total Caribbean	774	978.373	3,517	15,793	19,293
Total Central America	2,532	3,311.119	4,476	40,471	44,947
Total North America	42,028	40,466.66	716,753	55,936	772,689
Total North and Central America	44,560	43,777.779	721,229	96,407	817,636
Total Oceania	10,632	22,424.777	54,390	9,720	39,411
Total South America	135,428	96,291.822	224,888	173,316	398,134
Total World	536,480	1,531,858.6	2,028,400	1,326,283	3,329,897

the data. However, Matti Palo, an experienced researcher and analyst who has worked closely on FAO forest cover data has prepared a detailed repudiation[2] of these claims, and has forwarded these to Grainger: Grainger did not, he advises, use the two FAO surveys designed for the purpose of estimating deforestation, which show clearly that there has been considerable forest loss. Further, Palo argues that Grainger's suggestion that FAO has not taken natural forest expansion into account when assessing net forest losses is incorrect. Palo includes a number of other more technical criticisms of Grainger's methodology.

In our view, while there are questions in the global estimates as to the accuracy and completeness of the data on where deforestation is occurring; the level of secondary forest regrowth; and other forms of natural forest recovery, there is little doubt that significant losses of natural forest are occurring in the developing regions of the world.

3.2 Tropical Rainforests: A Key Concern

Most of the ten countries listed as the highest deforestation group in Table 3.2 earlier in this chapter contain some area of rainforest. Table 3.3 below shows figures produced by Hansen et al. (2008) using MODIS and Landsat imagery, which suggest that 43% of all global gross deforestation between 2000 and 2005 as estimated by FAO occurred in tropical rainforest.

Given these estimates, and the commonly recognized problems with sustainable forest management and subsequent degradation of forests in the tropics, it is understandable that many analysts and commentators interested in the broad climate change issue, have a preoccupation with tropical rainforests (Table 3.4).

Table 3.3 Ten countries with largest annual net forest loss, 2000–2005 (FAO 2006)

Country	Annual change ('000 ha per annum)
Brazil	−3,103
Indonesia	−1,871
Sudan	−589
Myanmar	−466
Zambia	−445
United Republic of Tanzania	−412
Nigeria	−410
Democratic Republic of the Congo	−319
Zimbabwe	−313
Venezuela	−288
Total	−8,216

---

[2] The authors have been provided with a copy of this communication by Dr.Palo; acknowledged here with thanks.

Table 3.4 Humid tropical forest loss, 2000–2005, millions hectare (Hansen et al. 2008)

	Area in 2005	Area lost 2000–2005	% lost 2000–2005
Global humid forest cover	1,152.5	27.2	2.4
Humid forest cover Brazil	312.2	11.2	3.6
Humid forest cover remainder of Americas	245.1	3.0	1.2
Humid forest cover in Indonesia	105.6	3.5	3.4
Humid forest cover remainder of Asia	318.1	8.5	2.7
Humid forest cover Africa	171.5	1.3	0.8

3.2.1 Forest Ecosystem Services

It is the case that the range of ecosystem services from forests, from carbon sequestration and storage to the less quantifiable values of biodiversity and other assets, is particularly important where tropical forests are concerned: they contain a disproportionately large share of the world's biodiversity, per unit area. Until recently, they have also been believed to have higher per unit of area biomass carbon content than most other forest types; however, a recent study by Keith et al. (2008) shows that average biomass carbon content on some of the sub-tropical and temperate forest sites they examined was significantly higher than that for tropical wet and tropical moist forests.

Forests and rain The United States National Aeronautics and Space Administration (NASA) has released a short video which encapsulates perfectly the powerful influence on global environmental sustainability that the world's rainforests have. In the space of a minute or so, the video plays a moving image of cloud generation and movement over the Earth's surface in a single year: Huge masses of cloud can be seen concentrating over the large remaining intact rainforest areas of the planet: South America, West Africa, and the Indonesian archipelago, up into mainland South East Asia. Large bands of cloud can be seen roiling off from these enormous formations over the rainforested areas, to parts of the globe as far away as southern Europe, southern and mid-western states of the USA, and the wet northern fringes of the world's driest inhabited continent, Australia. It becomes stunningly clear how dependent rainfall patterns in these areas (not to speak of within the rainforest nations and their nearer neighbours themselves) are upon these huge ecosystems.

These forests are global scale water sinks; rain that falls on them (itself a function, many believe, of the fact that the microclimates these forests create encourages that rain) is held in their humid, cool ecosystems, to evaporate slowly to form those clouds. It has been estimated (Global Canopy Programme 2007) that a rainforest tree can transpire eight to ten times as much water vapour into the atmosphere than an area equivalent to its crown cover in the ocean evaporates. If these forests were to disappear, rainfall over these areas would decline, and what did fall could create flash flooding, erosion and, as has happened in the Philippines in recent years, landslips on areas previously covered in forest.

This illustrates an important point that will be made throughout this report: While the global climate change issue – which so dominates world attention at the present time – and environmental sustainability more generally are moving together, there are some important priorities which must be assigned to the manner in which the global community approaches these issues. The NASA video shows that retaining rainforests will be of fundamental and increasing importance as the climate moves toward a generally warming and, in many parts of the world, drying environment. Given the speed with which rainforests have been and are being destroyed around the Earth, it may well be that humanity may not have to wait until the broad sweep of climate change itself produces these changes; they could happen well in advance of the overall warming trend.

Forests and international conflict There is also potential for international conflict over the consequences of declining rainforests as well. Many of the largest waterways on Earth draw heavily upon rainforests for their catchments, and flow through more than one country (the Amazon, the Mekong are examples); international political disagreements around these rivers are not uncommon, and could be aggravated by declining trends in their flows or water qualities.

In South-East Asia, the governments of Malaysia and Singapore have expressed deep dissatisfaction with the forests situation on the large Indonesian island of Sumatra, and in the provinces of Kalimantan, neighbouring the Malaysian states of Sabah and Sarawak on the island of Borneo. Their complaints arise from the effects of large fires there casting deep smoke palls over parts of Malaysia and Singapore, sometimes for weeks at a time. These fires are initiated by burning of logged over or partly cleared forests, which have gotten out of control in drier years, such as that arising from the major El Nino event in 1997. A consequence of these fires in Indonesia is that large areas of peatland are sustaining ongoing subterranean fires: peatlands in Indonesian are by far the richest source of terrestrial stored carbon, often containing five or ten times the carbon load of even the most heavily stocked rainforests. A reputational consequence for Indonesia is that it was often reported that during 1997, when huge fires raged across Kalimantan and Sumatra, GHG emissions from Indonesia were larger than the total carbon emissions for all of North America, and the idea that Indonesia is now among one of the top three or four GHG emitters has persisted, even though on a long term average basis it is not the case.

Forests and life Rainforests are also the repository of less tangible but in some respects deeper values to humanity in general: they contain large numbers of sites sacred to many people. Even the biodiversity issue (rainforests contain some 50% of all species of life on Earth), while often couched in scientific terms such as the importance of retaining gene pools and the delicate ecological balance that these complex systems themselves depend upon, also has more fundamental meaning to people who see the rich diversity of life on this planet as something of central importance in and of itself.

3.2.2 *Irreversibility of Rainforest Loss: A Key Concern*

It is the combination of these potential disasters with the *irreversibility* of rainforest loss that is most daunting. Rainforests are not like some piece of infrastructure which, if damaged by some natural or man-made event, can be repaired. They are not like an annual crop which, if lost to pest or weather, can be re-planted next year, and then genetically improved to be more robust in future. Their continued loss, therefore, constitutes an extremely high downside risk for humanity as a whole. This message is not captured in traditional economic and financial calculations, but should be kept in full view at all times.

3.3 The Implications of Reducing Deforestation

The question of the value of retaining natural forests is a global one; one that goes beyond the very real concerns about rainforests as outlined above, and which will occupy us for the remainder of this book: we will argue later that large increases in investment in sustaining and protecting forests will be needed to reduce deforestation globally, and we will consider the case for such investment. For now, we will attempt to establish that in terms of addressing the major global environmental issue we face – global warming – reducing deforestation is a significant factor. We stress that this is so regardless of the outcome of the debate outlined earlier in this chapter as to the overall net rate of forest loss; or the progress of the forest plantation sector. All we need to establish here is that there is a great deal of deforestation occurring in the natural forests – regardless of what offsetting regeneration or afforestation is occurring.

Much of the focus on the economics of sustainability in forests has included the search for ways to express in financial or economic terms all or some of the un-marketed ecological and other values that are contained within intact rainforest, and, given the recalcitrance of many governments in general, and their forest agencies in particular in the task of implementing sustainability, it is easy to understand what drives analysts to this search. However, as will be shown in Chapter 7, attempts to do this have run into the problem that assigning values to all of these goods and services have often resulted in astronomical aggregate figures that could not be completely monetized in any practical way, making such analyses prey to economists' responses that they serve no practical policy value.

Nevertheless, the success that many economists have had in debunking the case for sustainable management of forests does not justify the ethos that appears to permeate some economic and business circles, to the effect that if something cannot be monetized, then it can be assumed to be worth nothing: To a person who has plenty of air, that air is worth nothing: to a person who has too little, nothing could be worth more. It is difficult to conceptualize the amounts of money, and the order of risk that is present, in the areas of large scale forest loss we have presented in this chapter.

In this, we face the same dilemma in forests as Stern (2006) has identified more generally for climate change abatement strategies: the expected costs of these strategies rise rapidly as tipping points triggering major climate-related catastrophes approach. Because the occurrence of any given one of these cannot be predicted with accuracy, even as the weight of evidence unambiguously indicates significantly elevated levels of risk at the aggregate level, economic and political decision-makers will tend not to act until it is too late, or at least until the costs of acting effectively have become orders of magnitude larger than they are at present.

It will be argued throughout this book that there are far better (and cheaper) options available that could be taken now, rather than waiting on this disaster to unfold: In effect, if it is widely accepted that forest loss feeds directly into climate change, at globally significant levels, and that there is little disagreement now about the seriousness of this phenomenon for the future of humanity in general, then the ecosystem values of that forest related to climate change (and as will already be apparent from the forgoing discussion, this involves more than carbon) must be worth considerably more than nothing, and indeed, considerably more than is being invested in solutions globally at the present time.

References

Achard F, Eva H, Stibig H, Mayaux P, Gallego J, Timothy Richards T, Malingreau J (2002) Determination of deforestation rates of the world's humid tropical forests. Science 297(5583):999–1002

Fearnside P, Laurance W (2003) Comment on determination of deforestation rates of the world's humid tropical forests. Science 299:999–1003

FAO (2006) Global Forest Resource Assessment *2005,* FAO Forestry Paper 147. Rome

Global Canopy Program (2007June) Forests First in the Fight Against Climate Change. http://www.globalcanopy.org. Accessed 2009

Grainger A (2008) Difficulties in tracking the long-term global trend in tropical forest area. Proc Natl Acad Sci USA 105:818–823

Hansen et al (2005) 2000 vegetation continuous fields MOD 44B. In: Recent tree cover, collection 4. University of Maryland, College Park, MD

Lund HG (2000) Definitions of Forest, Deforestation, Afforestation and Reforestation. Available for download at: http://home.att.net/—gklund/DEF

Stern NH (2006) Stern review: The economics of climate change. HM Treasury, Government of the United Kingdom, UK

Williams M (2008) A new look at global forest histories of land clearing. Ann Rev 33(1):345–367

Part II
The Dynamics of Forest Loss

Chapter 4
Are Trade and Forests Survival Compatible?

Abstract This chapter examines the question of whether trade in forest products from developing countries adds or subtracts value from the standing forests in those countries, and what potential the trade instrument may have in promoting sustainable forest management in developing countries. It is important to bear in mind that only a few tropical countries export a significant proportion of their output, and that in general demand for forest products within developing and transitional economies has been growing rapidly. It is also the case that plantations are rapidly substituting for natural forest outputs in developing countries, not only for commodity grade material, but increasingly for higher quality raw material as well.

Forest products trade issues have come before the World Trade Organization (WTO) and international consultations in the United Nations Forum on Forests (UNFF) and the International Tropical Timber Organization. Voluntary certification of timber for compliance with sustainable forest management guidelines has grown quickly in the recent two decades, but has to date not produced significant price incentives for certified product from developing countries, nor major impact on deforestation. Illegal logging remains a major impediment to sustainable forest management in general in developing countries: the trade instrument will have some role in addressing this, but much more will be required for success.

This chapter examines the demand side, in the valuation of forest resources; we will discuss the forest valuation issue more generally in Chapter 7 of this book. Trade has many linkages with the future of forests and its impacts can be both negative (e.g. deforestation) and positive (e.g. generation of income and employment). The latter occurs when trade adds value to the forest resource, justifying its sustainable management. As the nexus between trade and forests is broader than what is caused by trade in timber and wood-based products, the interrelationship is very complex. It also explains why the question posed in the title to this chapter has generated major controversies.

In this chapter, we will attempt to clarify the key issues of forest trade, and examine how these could be addressed. Our examination will focus on how well international trade in forest products, especially as it has impacted upon the state of forests in developing countries, has served the valuation issue.

4.1 Where Trade Is Going: Emerging Trends

The gross value of forest sector production in 2007 was estimated at USD 950 billion of which value added accounts for USD 384 billion or almost 40% higher than in 2000[1]. The value of global trade of roundwood was USD 19 billion, wood products USD 114 billion, pulp and paper and paper products USD 196 billion and wooden furniture USD 54 billion. Trade has been growing faster than production in all types of wood-based products with the exception of pulp and particle board[2], and this trend is expected to continue albeit probably at a slower rate than during the last decade. Another major change has been the rapid growth in further processed wood and paper products.

About 80% of industrial roundwood is produced in developed countries. Only 7% of the world industrial roundwood production enters international trade but the share is about a third in sawnwood, wood-based panels and paper, and 23% in pulp[3]. On the other hand, the direct impact of trade on production is larger than these figures suggest; for example, in the case of tropical logs, sawnwood and plywood, exportable grades are selected; only a certain share will be taken, due to reasons of log quality[4]. This means that in order to produce 100 m^3 of exportable tropical sawnwood e.g. in Central Africa, 400–600 m^3 of logs may need to be felled.

The link between forests and trade is even more significant if indirect effects are taken into account. For instance in natural tropical forests, logging can open up new areas for encroachment through road construction making them accessible by shifting cultivators which can lead to deforestation. On the other hand, it needs to be emphasized that in the tropical countries which are worst hit by deforestation and poor management practices, the markets are essentially domestic and only a handful of developing countries are large scale exporters of forest products.

Trade has undergone major changes during the last two decades with changing patterns both in world exports and imports of forest products. In the past trade used to be dominated by three principal flows: (i) intra-regional trade between developed countries, (ii) raw materials exports from the South to the North, and (iii) finished products exports from the North to the South. This has however been changing as there has been a shift in labour intensive production from the North to the South (particularly in plywood, furniture and joinery products) serving international markets. Second, exports of pulp and reconstituted wood-based panel (mainly fibreboard) from the South have been expanding fast based on planted forests. The third significant factor is the increasing role of in-transit

---

[1] FAO (2008).

[2] The former is due to the importance of integration of pulp production with paper and paperboard production and the latter due to the importance of domestic market associated with the relatively low unit value and bulkiness of particle board.

[3] Calculated based on FAO Yearbook of Forest Products data for 2006.

[4] The exportable share of total production in sawnwood and plywood can vary from 20 to 80% depending on the product and species used and the quality control of the industrial operation.

countries like China and Vietnam which import roundwood and primary processed products from other countries in order to further process them into value added products for export to other countries. This trade has become significant and its impacts on the global trade flows will be felt over the next decade at least. Of particular importance is China's spectacular transition from a net importer into a major net exporter of furniture, joinery, plywood, panels and paper to the world markets, largely based on imported raw materials (logs and rough sawnwood). Due to its massive imports of wood raw materials, China and to a lesser extent some other in transit producer countries have been accused of having promoted deforestation in other countries.

This raises the question of who bears the responsibility for the impacts of trade on forests. In Europe, North America, Japan and some middle income countries, there is a growing recognition of trade's potentially harmful role on forests which has led to various initiatives to ensure that buying wood and paper products do not contribute to environmental destruction. Environmental requirements for forest management have become part of the demand function in major importing markets and particularly in some end-use segments (e.g. garden furniture, public construction works, printing papers, etc.). As a consequence, certification and legality requirements are now being applied as market standards in an attempt to provide assurance for buyers and consumers in developed countries that products they purchase originate from areas which are sustainably managed. Due to the complexity of supply chains involving different phases of processing and several successive transfers of ownership of products, improving transparency of raw material and product flows has proved to be a challenge, particularly in the case of in-transit countries. For example, imports of further processed products made of imported raw materials from China or other in-transit countries may not have utilized raw materials that have been certified in this manner.

The ultimate responsibility for legal compliance and sustainability of forest utilization relies naturally with the supplying country governments and their economic operators. However, this gets complicated as in many cases these operators are owned foreign companies and thereby serve the interests of parent companies.

Demand for forest products is now growing faster in the South and countries in transition (particularly Russia), while markets in the developed countries are mature with only low growth rates or even decreases in some cases. In some products like newsprint, demand is generally expected to decline in the developed countries due to loss of the market share of printed media. The global economic crisis which was current at the time of writing is likely to induce structural changes in consumption patterns which may lead to slower long term growth in paper and paperboard products. On the other hand, wood's credentials as an environmentally friendly building material are increasingly being recognised, and this could lead to increased demand in this sector. Another factor is the fast growing demand for wood-based fuels, opening up markets for low grade materials which could not been previously used. In the long run, trade in this commodity can be expected to take place in liquid form but in the short to medium term wood fuels are likely to be traded in increasing volumes in chips or pellets.

Growing demand in the South will boost intra-regional trade, particularly in Asia and Latin America where internationally competitive forest industries have developed during the last two decades. In Africa the situation is more difficult as the deficient infrastructure in the interior of the continent is a major constraint for trade development between neighbouring countries.

4.2 How Future Demand Can Be Met: Rapidly Changing Supply Patterns

4.2.1 The Raw Material Base

The raw material base is also changing as a result of the increasing role of plantation wood. In 2000, plantations were 5% of the global forest cover, but they provided some 35% of harvested roundwood, an amount anticipated to increase substantially in the future (FAO 2006). In some countries, particularly in Latin America, plantation development has benefited from subsidies which undermine the value of competing wood production from natural forests. Subsidies have been justified on varying grounds including the socio-economic benefits of plantation-based industrial development and difficulties in expanding production based on natural forests. This has coincided with the mounting constraints on market access to timber and timber products from natural tropical forests due to concerns related to legal compliance and sustainability of forest management. For tropical timber producers over the last two decades, maintaining competitiveness in the main import markets has mostly been an uphill battle and it is not certain that this can be changed by certification or regulatory means.

Plantation timber and timber from natural tropical forests are not full substitutes. Much of the international trade of tropical timber from natural forests is in high-value species, while plantation wood mainly goes to commodity products like pulp, reconstituted wood-based panels (fibreboard and particle board). This means that the decline in topical timber trade has not to date been overtaken entirely by plantation timbers, but primarily by temperate species such as oak, beech, and others produced in North America and Europe. However, this situation is evolving. For example, plantation wood such as pine, eucalyptus and acacia has been traditionally used for bulk products like pulp and reconstituted panels, fibreboard or particle board. More recently, the end uses have expanded to solid wood products like lumber and plywood in which these plantation species can to a certain extent replace timber products derived from natural tropical forests. On the other hand, there is now a growing interest in establishing plantations of high value species such as teak, mahogany, cedar, and others, which would in the long run be direct substitutes for timbers from natural tropical forests. The substitution relationships are therefore complex and they are influenced by the competitive advantages of plantations such as higher returns, faster growth rates and less controversy compared to production based on natural forests.

4.2.2 Forest Industry

Forest industry is still a relatively fragmented sector compared to e.g. manufacturing of steel, plastics or chemicals which are wood's main substitutes. Concentration of industrial production is nevertheless progressing, not least because concentration is also taking place among customer industries. As an example, in timber products markets large furniture companies (such as IKEA), DIY chains and builders' woodwork retailers (such as Home Depot, Kingfisher/B&Q, Lapeyre, Leroy Merlin, etc.) are operating globally or regionally, and they have become increasingly important market drivers with specific purchasing requirements related to timber and timber products. The process of industry concentration is not limited to developed countries but it is also taking place in the South where new global-level players have emerged in pulp and paper products (e.g. Sappi and Mondi from the Republic of South Africa, APRIL and Asian Pulp and Paper in Southeast Asia).

In the late 1990s forestry and forest-product industries provided 47 million jobs worldwide. Of those, forestry, wood industries, and furniture activities each generated 10–15 million jobs[5]. Practically 70% of the employment in forestry, wood industries, and furniture came from informal and subsistence activities. Particularly in the tropical countries most of the forest industrial employment is generated by small and medium-sized enterprises (SMEs) many of which are run by community organizations or cooperatives.

The role of the small and medium-scale enterprises (SME) has traditionally been strong in the forest sector, but poorly recorded. The sector has traditionally focused on local and domestic markets not being able to compete internationally apart from niche markets. It has been estimated that about 70% of the employment in forestry, wood industries, and furniture came from informal and subsistence activities. Forest-based SMEs typically serve local or domestic markets being run by owners themselves with hands-on knowledge on the operations, rather than relying on formal management systems. Only a few of these enterprises tend to develop into medium-sized or larger enterprises which apply modern technology and management methods. Especially in the tropical countries most of the forest industrial employment is generated by SMEs many of which are run by community organizations or cooperatives (ITTO 2007). Managerial weaknesses are common among them due to lack of qualified human resources and inadequate management and information systems. This results in low efficiency and difficulties for certification and tracing of products, thereby limiting access to the markets which demand proofs of legality and sustainability.

While tropical timber industries, in relative terms, should enjoy competitive raw material costs, several factors make industrial wood expensive such as increased management costs due to legal and certification requirements, poor planning and outdated equipment in harvesting resulting in wasteful practices, inadequate

---

[5] In addition, about five million people worked in the pulp and paper industry.

infrastructure, and high transaction costs. Therefore the economic incentives for illegal operations which do not incur these costs tend to be significant. This works against forest management units and industrial enterprises which are complying with regulations and voluntary sustainability standards. Furthermore, in spite of the high raw material costs, there is a high rate of waste further down the value chain. For example, in Bolivia in a fully integrated production operation only 15% of the timber volume felled may end up in the value added products while about a half of the felled volume remains in the forest and the remaining 25% is industrial processing residue which can be only used as fuel.

The supply side in the forestry sector is also influenced by increasing demands for recognition of the tenure rights on forests by indigenous and other local people and their communities: we will return to this subject in Chapter 7 of this book. There is some evidence that in countries like Mexico, India, Nepal, and others community forests can be sustainably managed by local people, and these forests can become an important organized source of supply of forest products and services. However, creating exportable production units among these new market actors is a formidable task.

In our view, the world's forests will be able to meet the future demand for forest products needs but there will be significant changes in the future supply chains will be different from those of today with more wood coming from plantations, more products supplied by developing countries to the world markets, and more environmental services becoming subjects of trade. However, it is uncertain to what extent community-based and other SMEs can exploit opportunities offered by markets requiring proof of legal compliance or sustainability as they are already disadvantaged in the international marketplace compared to large-scale commercial enterprises.

4.3 Are Impacts of Trade Liberalization on Forests Positive?

The World Trade Organization WTO Agreements define the international trade rules. Among those affecting forest products trade are the General Agreement on Tariffs and Trade and Agreements on the Technical Barriers to Trade (TBT), the Sanitary and Phytosanitary Measures (SPS), the Subsidies, the Trade Related Intellectual Property Rights. The WTO rules are based on the principles of non-discrimination and equal treatment of like products. The TBT agreement lays down how standards are developed and applied, how phytosanitary rules can be used to avoid entry of pests and diseases, how specific trade related taxes can be imposed or fiscal subsidies provided, or how the property rights of traditional forest related knowledge be protected. The WTO recognizes in its Marrakech Agreement of 1994 that expanded trade should allow optimal use of the world's resources in accordance with the objective of sustainable development. This is an expression of the trade regime moving beyond tariff reduction and non-discrimination towards new ways to combine social and environmental protection with a continuing commitment to liberal trade, while allowing for domestic government intervention within specified common rules.

As a result of the application of international trade agreements in developed countries, tariffs on primary forest products are generally very low or even nil but there is still a significant element of tariff escalation along the value-added gradient, i.e. higher tariffs for value added products than for logs, sawnwood and standard grade wood-based panels. The situation is different in developing countries where all tariffs tend to be significantly higher than in developed countries and tariff escalation can be much stronger. This in general makes imports unduly expensive for consumers and it seriously undermines the development of intraregional trade.

From the trade policy perspective it is important to clarify the impact on further trade liberalization on forests, which requires analysis of direct and underlying causes. Due to relatively low levels of tariffs in wood raw materials and primary processed products in the traditional major import markets of developed countries, the scope of further trade liberalization is limited. However, by reducing the prevailing tariff escalation, several important benefits could be expected for developing countries. Growth in further processed products would be increased to create value in the countries where forests are located. Another positive impact for consumer gain would be increased intraregional trade among developing countries at lower costs than at present.

4.3.1 Winners and Losers Under Trade Liberalization

The economic benefits from forest products trade have been poorly distributed both within and between countries. Those with inadequate infrastructure, inefficient bureaucracies, weak legal regimes and poor macroeconomic stability will not be able to increase their exports. There are divergent views among stakeholders, with some NGOs being against trade liberalization (along with some governments wishing to protect domestic industries) while importing countries seek lower costs. An independent assessment of impacts on forests of further trade liberalization (Katila and Simula 2005) revealed that the overall economic gains would be positive but globally not very large. This is due to the fact that the current level of import duties is already fairly low in major import markets and therefore the impact would depend on the extent to which trade would increase among developing countries. Forest product consumption and production in aggregate may not be influenced significantly but some countries would benefit from increased trade and in their case the economic value of forests would increase.

Net global roundwood production is predicted to increase only by about 0.5%. Trade liberalization would open opportunities for the forest resource-rich countries, such as Brazil, Indonesia, Malaysia and China (except that China's huge domestic demand will limit export opportunities) and also some West and Central African countries. Consumers in the importing countries would gain in terms of welfare because they would face lower prices as a result of liberalization. Because tariff escalation for forest products is the highest in South Asia and Middle East and North Africa, they would also experience largest consumer gains under conditions designed to improve trade.

On the other hand, the environmental and social impacts of trade liberalization would be ambiguous. In countries with governance problems the impacts could be negative and if these coincide with biodiversity hotspots (such as those Brazil, Papua-New Guinea, Indonesia, the Congo Basin) the impacts could be irreversible. Developing countries with forest industries which are protected by high tariffs may incur considerable social costs due to gradual downsizing of industrial capacity even though many companies may still carry on inefficiently, with limited social contribution. If processing capacity exceeds the sustainable production potential of the forest, as is the case in many parts of the tropics, downsizing would be positive from the environmental perspective: it is possible that social costs of trade liberalization might outweigh economic gains. Therefore it is important that trade proceeds on the basis of adoption of adequate safeguards through a precautionary approach (which might entail using a phased approach to trade liberalization). In future trade agreements, be they multilateral or bilateral, specific mitigation and enhancement measures will be needed to make liberalization work towards sustainable forest management.

4.4 Can Trade Rules Differentiate Sustainably Produced Forest Products?

While import tariffs have been declining, non-tariff barriers are increasingly impacting upon trade in forest products. Technical standards and regulations for forest products and how they are produced are regulated under the WTO Agreement on Technical Barriers to Trade. As forest management deals with a natural resource in a local social context and with environmental impacts it has proved to be difficult to interpret the TBT Agreement in the case of forest products. Voluntary standards such as those applied in forest certification and mandatory technical regulations have impacted upon the competitiveness of developing countries. For example, the EU Construction Products Directive (CE marking of products) which was issued to ensure that materials have adequate strength properties for construction has been problematic as it requires testing by independent accredited laboratories which do not exist in many developing countries; nor do many of them have independent national accreditation bodies.

Forest certification and labelling are permitted under the jurisdiction of the TBT Agreement provided they are voluntary and do not lead to discrimination of like products. However, any mandatory requirements related to marking or labelling based on origin of forest products have been determined to be against the trade rules, and countries which have imposed them have had to cancel such regulations. A number of initiatives have been taken at country and international level to make voluntary forest certification mandatory, but this would mean that the TBT rules designed to prevent these instruments becoming a source of discrimination would then apply to them. Mainstreaming of forest certification standards in national regulation has been implemented by some countries through different incentives (e.g. Bolivia, Brazil, Peru, the Republic of South Africa) but in principle forest

4.4 Can Trade Rules Differentiate Sustainably Produced Forest Products? 63

certification remains voluntary. However in practice it has become a precondition of access to many markets and therefore the extent to which it is in fact voluntary can be questioned. Since sustainable forest management can be achieved only in the long run, and since also it has been understood that certification is difficult for developing country producers, a phased approach has been adopted in many market requirements, obliging suppliers to provide proof of legality of their forest products as the first step towards sustainability.

Government timber procurement policies are being implemented in about ten countries in order to ensure that wood-based products (including paper) come from legal and sustainable sources, often allowing a fixed time horizon for delivery of proof of sustainability. Public procurement policies are regulated by the Plurilateral Government Procurement Agreement (GPA) which has been ratified by fewer that 40 WTO members, which limits its area of application. The timber procurement policies tend to specify forest certification as one of the possible proofs for sustainability and legality provided that the respective certification systems meet a predetermined set of minimum criteria.

Trade restrictions resulting from any Multilateral Environmental Agreement (MEA) have not yet been challenged in the WTO. Were there a challenge, trade restrictions on environmental grounds could be ruled in breach of international trade law. Usually the most recent international instrument always applies if there is conflict between a MEA (e.g. possible trade restriction measures to protect biodiversity under the Convention on Biological Diversity) and the WTO agreements. As the latter are relatively recent, they would probably prevail in many cases but there is no certainty about this.

Article XX under The General Agreement on Tariffs and Trade (GATT) – the predecessor to the World Trade Organization – allows taking national measures which are necessary to protect human, animal or plant life or health which can be interpreted relevant for forest protection. On the other hand, GATT prohibits discrimination of 'like products' based on production and processing methods (PPMs). Trade discrimination for unsustainably produced forest products is not possible without applying it to all imports and domestic production if the measure does not qualify under GATT Article XX. According to the WTO, the market access should be equal for 'like products' but this is not currently the case because of existing central-government public procurement policies.

4.4.1 Protection of Forest-Related Intellectual Property Rights

Closely related to the Convention on Biological Diversity (CBD), the WTO Agreement on Trade-Related Intellectual Property Rights (TRIPS) provides protection of traditional forest related knowledge and attempts to balance private and public interests. Micro-organisms and microbiological processes can be patented to promote their development but plants and animals are excluded. However, the boundaries are not clear; for example, protection of plant varieties under the TRIPS Agreement is controversial. Key issues from the forest perspective have been access to, and sharing of, benefits from forest genetic resources. Patenting is however not suitable for the protection of forest-related traditional knowledge and it is an open question as to how its utilization by outsiders should be compensated if it leads to economic benefits for them. Obviously, the current situation does not contribute to the valuation of the forest resources as owners can only benefit from their intellectual property rights apart from voluntary agreements with outsiders (such as pharmaceutical companies) but they are rather an exception than a rule.

4.4.2 Trade in Intergovernmental Forest Agreements

The recent Non-Legally Binding Instrument on All Types of Forests (NLBI) negotiated under the auspices of the United Nations Forum on Forests (UNFF) was approved by the Economic and Social Council of the United Nations in December 2007. The purpose of this instrument is:

- To strengthen political commitment and action at all levels to implement effectively sustainable management of all types of forests and to achieve the shared global objectives on forests
- To enhance the contribution of forests to the achievement of the internationally agreed development goals, including the Millennium Development Goals, in particular with respect to poverty eradication and environmental sustainability
- To provide a framework for national action and international cooperation

It was the outcome of more than 10 years of intergovernmental negotiations on forests (we discuss these in Chapter 6). The NLBI includes four Global Objectives on Forests and provisions for their achievement through national measures and international cooperation. The Instrument does present a common vision on how sustainable management of the world's forests might be achieved. However, the negotiation process trade issues (together with financing) proved to be particularly controversial and the agreement does not contain any clause for trade restrictions.

Some observers have suggested that aspects of forest products trade might be addressed under the United Nations Convention on Biodiversity (since, as we will discuss in Chapter 6, there is no equivalent convention at this level for forests). To date, however, it should be noted that the CBD has not engaged with the forest products trade issue: it focus has been on access to and sharing of benefits from sustainable utilization of biodiversity.

The International Tropical Timber Agreement (discussed in Chapter 6) renegotiated in 2006, defines its objectives as:

> …to promote the expansion and diversification of international trade in tropical timber from sustainably managed and legally harvested forests and to promote the sustainable management of tropical timber producing forests[6].

---

[6] Article 1 of the Agreement.

However, the implementation body, the International Tropical Timber Organization (ITTO) has been ineffective in this area, because the Agreement does not attempt to regulate trade effectively, focusing instead on promotion of tropical timber trade from legal and sustainable sources. In fact, it is explicitly stated in the Agreement that:

> Nothing in this Agreement authorizes the use of measures to restrict or ban international trade in, and in particular as they concern imports of, and utilization of, timber and timber products."[7]

Within its constraints in trade regulation, ITTO's promotional role could be substantially enhanced in the area of reducing effective barriers to market access and promotion of markets and marketing of legally harvested and sustainably produced tropical timber and timber products. ITTO's dilemma comes from the fact that consumer member countries prioritize conservation, sustainability and legality, while producing member countries focus on economic development as being most important.

In the late 1990s there was an initiative to conclude a free trade agreement on forest products driven by the United States and strongly supported by the private sector with an interest in export expansion. Some major environmental NGOs fiercely lobbied against such an agreement, arguing that it would lead to increased logging and deforestation. After the Seattle WTO Ministerial Conference in 1999, the idea was shelved (due in part to the dominance of other issues of the Doha Development Agenda such as agricultural subsidies). The United States therefore proceeded with the idea through bilateral trade agreements and promotion of the Free Trade Area of the Americas (FTAA). The stalling of the WTO Doha Round has encouraged other trading partners to proceed with bilateral free trade agreements, even though these appear to which appear to yield fewer benefits from forest products trade to developing countries than a multilateral solution. It is worthwhile to noting that new US trade agreements involve an environmental assessment.

4.4.3 Pressures from Non-governmental Organizations

Some NGOs have called for amending trade law to specifically allow different treatment of sustainably and unsustainably produced timber. Although similar issues do not (yet) appear to be of concern for the other manufacturing industries, they may emerge in the future. For instance, extensive charcoal use from natural forests in the Brazilian steel industry is already an important environmental concern in the country but there is apparently little awareness among international steel buyers of the issue. Biofuels are already heavily criticized for their possible negative impact on forests and the environment, as a result of increased pressure on resources in the tropics. Increasing environmental and social demands (particularly those related to climate

---

[7] Article 34 of the Agreement.

change) are gradually spreading beyond wood-based products and this could contribute to a level playing field with regard to substitutes, eventually.

Civil society initiatives to influence forest products trade have been driven by developed country NGOs, but consumer awareness on the issue is still generally low. NGOs working to influence forest product suppliers have mainly focused their campaigning against large corporations with headquarters in developed countries, as these are relatively easy targets due to their exposure to large clients with a direct consumer interface. The challenge is more difficult in the case of corporations based in developing countries; they depend less on the environmentally and socially sensitive markets in developed countries where international NGOs have their primary influence. Companies in developing countries have fewer incentives and often less interest in engaging in social responsibility or environmental conservation approaches such as certification, as long as their clients are not expressing specific requirements or concerns. This is probably why NGOs have partly shifted their emphasis to lobby for trade regulation which would prevent illegal products entering the developed country markets. We will take up this issue again in Chapter 6.

Consumer concerns appear to be more directed towards climate change than forest sustainability and legality. Carbon footprint or other similar indicators related to climate change are gaining increasing attention in campaigning for environmental conservation. On the other hand, emerging public procurement policies for forest products and green building codes have established a new situation in which climate change, legal compliance and forest sustainability are considered in an integrated manner as a market requirement. This is fully in line with the effectiveness objective of these policy tools.

4.4.4 Taking Stock

Trade rules can allow differentiation of sustainably or legally produced products but this cannot yet effectively linked to their origin. Domestic products should be treated in the same way as imported ones to avoid disguised discrimination and the treatment of tropical timber needs to be similar to that of timber from temperate or boreal forests. GATT Article XX allows some flexibility if a measure is justified based on a need to protect human, animal or plant life or health. The current approach relies on the market to solve the issue as it is unlikely that sector-specific requirements under WTO could be agreed upon. Climate-friendly products such as timber could benefit from a demand boost, particularly in applications like building and construction. The available life-cycle assessments (LCA) between materials (Athena Institute Forintek, 2009) have clearly demonstrated climate benefits of wood against substituting products and recent research also suggests that wood-based fuels would be more efficient than alternative biofuels as substitutes for fossil energy sources (FAO 2008 op cit). In spite of these promising indications, due to ambiguities in making LCA-based comparisons between materials, it is uncertain whether wood-based products could be broadly accepted as environmentally friendly products and treated in trade on more favourable terms than their competitors.

As trade regulation in forest products has proved to be controversial for creating economic value and protection of environmental and social functions of forests, there has been an emphasis on voluntary measures which are outside the scope of intergovernmental agreements. Many large private companies involved in trading wood-based products (furniture companies, DIY groups, etc.) and companies in publishing and packaging which are exposed to NGO pressure in developed countries have adopted private purchasing policies which are aimed at mitigating corporate image risks related to their supplies of forest products. These policies usually make use of voluntary certification schemes as a tool to ensure that the products marketed by them are not perceived by consumers as contributing to deforestation or forest degradation.

4.5 Has Certification of Forest Management Created Value for Forest Resources?

In Chapter 2, we made the point that addressing issues such as forest certification, and illegal logging (discussed next in this chapter) in isolation is unlikely to generate significant progress with sustainable forest management. Rather, all such measures will need to be placed in the context of larger economic and social developments to forest outcomes; the following discussion of the specifics of certification and illegal logging should be seen in this light.

The idea of forest certification was introduced almost about 20 years ago as a 'positive' measure to combat deforestation. This outcome has not materialized for tropical forests in particular. In our view this is because the underlying reasons for deforestation are only partly related to trade in forest products, and the role of export trade is in any event quite limited for most developing countries. However, as explained earlier, there are factors (such as logging-associated road construction which opens up new areas for encroachment (Chomitz and Gray 2006) where certification has a role: due to weaknesses in regulation and enforcement. More recently it has been proposed that private forest certification schemes could be a means of regulating the market by fostering the voluntary adoption of sustainability standards for forest management through market dynamics (Cashore 2002). This is in line with the overall tendency to increase the role of non-governmental instruments towards achieving environmental and social goals.

4.5.1 Lagging Developing Countries and Uncertain Market Benefits

Developing countries continue to lag behind developed countries in achieving forest certification. In 2008, about 320 million hectares of forests were certified worldwide (UNECE/FAO 2008). Of the total area, developing countries accounted only for 7%

in 2007, or about the same percentage as in 2002 (Purbawiyatna and Simula 2008). Developing countries produced 27% of world industrial roundwood production in 2007, which was almost four times higher than their share of the world's certified forests. In total, no more than 8.3% of the world's forests are certified. Nevertheless, certification has become a mainstream activity in many developed countries, but only in few developing countries (e.g. Brazil, Chile, Malaysia).

In 2008, the potential roundwood supply from the world's certified forests was approximately 416 million m^3, which is one-quarter of the world's total industrial roundwood supply (UNECE/FAO op cit). Most of this timber is sold without label or reference to certification, due (in part at least) to the fact that in major consuming countries there has been little real demand for certification of domestically produced timber. Concerns tend to apply only to imported products, regardless of where they come from.

However, global demand for products from certified forests is growing. In some key European markets demand is already significant, although the volume is unknown. The key drivers of certification are public procurement policies as well as business-to-business demand supported by corporate social responsibility and sustainability initiatives in the building and construction sector. Green building codes are increasingly influencing the market for wood used for construction. Many buyers are committed to procuring only legally sourced timber, giving preference to products from sustainable sources, and have long-term policies to obtain all supplies from such sources. In some markets and market segments, demand for certified timber exceeds supply, particularly in the case of hardwood products.

Price premiums would be required to pay for the additional costs of certification audits and achieving compliance in forest management with certification criteria. Market studies have shown that there is some willingness among buyers to pay a higher price for certified products but it is far from being widespread. Price and product characteristics continue to be the main purchasing decision criteria among consumers and industrial buyers. Mainstreaming of certification means that, in the long run, the market is unlikely to pay a premium, particularly as market access to uncertified products is gradually becoming more and more limited anyway. Certification is already becoming a prerequisite for access to many market segments in developed countries. With the exception of niche markets, internalizing the costs of better forest management through certification is therefore occurring mainly through access to markets rather than through price premiums. This is unfortunate as certification should add value to forest products as a result of the information that it provides to buyers and consumers on the quality of forest management where harvesting has taken place.

4.5.2 *A Tug-of-War Between International Schemes*

At the global level, there are two competing certification schemes with different operating modalities (Box 4.1). The Forest Stewardship Council (FSC) provides

> **Box 4.1** Main characteristics of internationally operating forest certification systems
>
> Forest Stewardship Council (FSC) established in 1993
> - One comprehensive global system
> - Develops national standards but in their absence draws on the generic Principles and Criteria for responsible forest stewardship
> - Does not recognize any national certification systems
> - Centralized accreditation of independent certification bodies
> - Applies a common label
> - Governance based on three chamber system (economic, social and environmental) with a balance between the North and South
>
> Programme for Endorsement of Forest Certification (PEFC) established in 1999
> - Mutual recognition mechanism for national certification systems
> - Defines specific requirements for standard setting, standard contents, and procedures for certification and accreditation
> - Applies a structured procedure for assessment and approval of applicant national systems
> - National accreditation of independent certification bodies
> - Applies common label
> - Governance based on representation of national systems and stakeholders

all the necessary elements of certification through centralized decision-making on standards and accreditation. The Programme for the Endorsement of Forest Certification (PEFC), on the other hand, operates as a system for mutual recognition between national certification systems (operating or under development in 32 countries). In 2007 almost two-thirds (65%) of the world's certified forests (in 22 countries) carried a PEFC certificate, while the FSC's share is 28% (in 78 countries); the remaining forests are certified solely under national systems.

Certification continues to be one of the most controversial issues in the forest sector in spite of the fact that competing systems have been evolving towards compatibility and convergence. In practical terms, the many similarities between certification schemes should offer a basis for cooperation but there still continues to be strong competition between the supporters of FSC and PEFC. This arises from some important differences between these schemes which are embedded in the deep values of their supporters. These differences are related to three main aspects: the process of developing forest management standards; some aspects in the contents of these standards; and the overall governance arrangements.

The differences are not so much technical as essentially political. Crucial differences in standard setting between schemes appear to be related to: (i) the meaningfulness or effectiveness of participation by interested parties; (ii) interpretation of situations in which a stakeholder group does not participate even though it is invited to do so;

and (iii) the possible dominance of certain parties. These three aspects are all considered to be important elements of credibility.

Despite the differences present, standard-setting processes under various certification systems have had a positive impact on stakeholder participation in all the countries where national standards have been developed. However, it has remained difficult for the majority of forest owners and forest industries to accept that one single group (economic, social or environmental) can effectively veto the conclusion of a draft standard which has been developed through a participatory and transparent process. In the same way, refusing to participate in the standard setting process is used as a means to discredit the result independently from the outcome. Such an absence can then be easily associated with dominance of certain parties in the standard setting process independently from the participatory rules that have been adopted.

There are several cases where standard setting processes at national level have been very extended processes, taking several years, due to the fact that the parties have not been able to reach an agreement which is acceptable to everybody. This is obviously very difficult in a subject like forest management where perceptions, values and interests among stakeholders tend to differ. On the other hand, some of the parties involved have not adequately understood that moving to a higher level of performance in forest management is a stepwise process which can be achieved through periodic adjustments of the standard requirements – an inherent feature of all the existing forest certification systems. This is a classic case of the perfect being allowed to get in the way of the good.

There are also some important issues related to the substantive requirements of forest certification standards. For example, certification of planted forests which are being expanded at a rapid rate in many developing countries has become controversial even though several millions of hectares have successfully passed sustainability audits. Another controversial aspect has been how forests of special conservation value should be treated in forest management, i.e. which areas should be put aside for biodiversity protection and how other such valuable forests should be utilized. It is important to note that the applicability of any global forest management standards is problematic as the world's forests vary widely in terms of their bio-physical characteristics, socio-economic contexts and ecological situation. We are strongly of the view that global standards should only provide a framework within which national or local standards can be developed.

4.5.3 Improving Effectiveness of Certification

The above issues have also economic implications. In addition to influencing market access, certification can also significantly impact upon costs, particularly in developing countries where operators are poorly equipped to implement information and management systems which can make forest management auditable. In developed countries the costs tend to be marginal as forests are already relatively well-managed and information on resources and activities carried out is well recorded. Losers therefore tend to be smallholders, community forests and in general producers of timber from natural tropical forests.

In this sense, certification runs a risk of becoming a perverse instrument, promoted with a vision to prevent deforestation but in fact placing developing countries and most of their operators at a disadvantage. Only through massive support to developing countries could such perverse impacts be mitigated and, as in so many other issues in forests, this support has not been forthcoming. There are some public and private sector initiatives to this end but they are grossly inadequate for the actual needs. If the barrier to achieving market acceptance of products through certification is set too high, it can serve as an excuse to continue exploitive heavy logging, targeting markets which are indifferent to sustainability concerns, or even to convert forests into other land uses.

Despite the problems and issues discussed above, forest certification has the potential to promote sustainable forest management and thereby to some to extent internalize its costs in forest products markets. It underscores the importance of well-ordered trade in wood and other forest products in compliance with relevant national and international laws and voluntary standards concerning how stakeholders perceive sustainability in practice. The accumulated experience of the forest certification approach has produced valuable lessons which can be used to verify an environmental service such as carbon sequestration, biodiversity conservation and watershed protection. In all these cases, remuneration of the maintenance or enhancement of an environmental service to forest owners and managers would have to be based on verification. Certification of sustainable forest management provides a comprehensive approach to this purpose covering the economic, social and environmental pillars of sustainability.

Progress with certification could be accelerated if the contention between supporters of the two international certification schemes could be contained. All the parties involved should assess what has been gained from certification, and what opportunities have been lost due to continuous conflict. It is high time that all involved accepted that no single certification system can be applied in forestry. Healthy competition between schemes is important for their continuous improvement. Certification is a management tool for improved forest management; the ideological conflicts between stakeholders should be considered in for a where they can be effectively addressed, rather than played out as random skirmishes in the field and the literature on this subject.

4.6 A Distorted Playing Field: Addressing Illegal Logging

Illegal logging and associated trade is a cause of serious concern, not only because of linkages with corruption and crime but also as it affects valuation of forests, thereby undermining any efforts to achieve sustainability in forest management. Illegal forest activities contribute to the degradation of forests and undermine the contribution of the forest sector to employment generation, social and economic development and poverty alleviation. An unknown volume of timber is illegally felled, processed and traded but it has been estimated that in some countries illegal harvests have exceeded those that are officially sanctioned. The World Bank (2006) has estimated that US $10 billion is annually lost in assets and revenue due to illegal logging on public lands in developing countries and another US $5 billion is lost fiscal revenue to governments due to evaded taxes and royalties. In addition, illicit operations under-value the resource resulting in forest degradation and suboptimal socioeconomic benefits for local people. Were illegal logging eliminated or reduced to marginal levels, world prices of industrial roundwood are estimated to increase by 1.5–3% and 0.5–2% for processed products (Li et al. 2008)

There has been no level playing field for responsible legal operators in the international market. The problem of illicit activities is mainly confined to developing countries and some countries with their economies in transition. As we have noted earlier in this chapter, timber trade involves complex trade flows often including trans-shipments and further processing in transit countries. Many loopholes exist for falsifying information and documentation. There is a common perception that illegal operations are widespread particularly in the tropical timber sector. Increasing awareness of the problem of illegal logging and illegal trade in forest products has tainted the image of the whole forest sector which has led major importing countries to introduce requirements for legality to provide assurance for buyers and consumers of wood-based products that they are not contributing to criminality in exporting countries through their purchasing. In some importing countries it has become morally unacceptable to use illegal timber products in public procurement. It is also broadly recognized that trade is an essential element in addressing forest law enforcement and governance.

4.6.1 Causes of Illegal Logging and trade

Several analyses on the problems leading to illegal logging have underscored weak governance and enforcement as well as the contributing role of timber trade and markets (see ITTO/FAO, 2005; Contreras Hermosilla et al. 2007; Kaimowitz 2003; Tacconi et al. 2003; World Bank. 2006 op cit). The situation varies by country depending on the political institutional and legal context, forest management system, forest product markets, social factors and traditions that apply. The recent analysis by ITTO and FAO (ITTO/FAO 2005 op cit) found five general factors contributing to illegal activities in the forest sector:

1. *Flawed policy and legal framework* which results in distorted economic incentives promoting illicit operations. Legislation may be incoherent, unrealistic and not enforceable leaving important caveats undefined such as forest land tenure and use rights. Due to excessive regulations, transaction costs of legal operations become prohibitively high and are perceived as unfair, making it impractical or impossible to respect the law. This is particularly the case for community forests and SMEs which are ill equipped to comply with extensive documentary procedures and may therefore appear as operating outside the law. Where there is widespread illegal activity in forests, there is often unclear or inadequately defined forest tenure, weak political institutions, skewed distribution of income, inadequate

planning, poor information and monitoring systems and generally lax law enforcement. There is lack of understanding of the full impacts of illegal operations for socio-economic development, fiscal revenue, and environmental conservation among policy makers to push them to make necessary policy reforms.
2. *Minimal enforcement capacity* is due to institutional weaknesses. Powerful political interests directly or indirectly involved in illegal forest operations makes enforcement ineffective. With a lack of alternative economic opportunities for local people, there is not enough public pressure to tackle illegal forest activities leading to their silent acceptance. Lack of intra and inter-agency coordination between enforcement and judicial bodies makes the risk of being caught and prosecuted low contributing to economic incentives driving illegal operations.
3. *Insufficient data* on the condition of forest resources and their change over time, production activities, illegal operations, timber flows within the country, volume of cross-border and other trade due to insufficient statistical systems, and markets makes it difficult to reliably monitor what is happening in the forests and the supply chain.
4. *Corruption* in government and the private sector is possible due to a lack of transparency in policy implementation, marginalization of the rural poor and lack of public pressure to tackle the problem. Of particular concern is the corruption related to allocation of forest concessions and use rights and control of forest harvesting and timber transportation for avoiding payment of taxes and fees. Enforcement of regulation of industrial capacity is flawed, leading to excessive demand for logs driving illegal forest land conversion, logging in national parks and other conservation areas, and unsustainable harvesting in production forests.
5. *Market price distortions for wood products* prevail in domestic and export markets which are indifferent to legality and sustainability of sources from where tropical timber and timber products originate. This is due to the existence of ready outlets for low-priced illegally harvested products. In problem countries producers have inadequate incentives to demonstrate legality and sustainability of their operations as they cannot compete price-wise with illegal operators. Uncertainty over expected tangible benefits as well as constantly changing goalposts of different international market requirements related to legality and sustainability make it difficult for producers to take systematic corrective action, and the long-term investment needed to support this.

4.6.2 Emerging Market Requirements

As a response to this situation initiatives have been taken at the international level or importing countries such as e.g. the regional Forest Law Enforcement and Governance (FLEG) processes coordinated by the World Bank and the EU Forest Law Enforcement Governing Trade (FLEGT) Action Plan which provides for voluntary partnership agreements (VPAs) with supplier countries to ensure that only legal timber can enter the EU market. Partner countries are expected to implement a timber licensing scheme and EU border control agencies allow imports from these countries only if they are accompanied by FLEGT licenses. At the end of 2008 one VPA has been signed (with Ghana) and formal negotiations were under way with four other countries.

As the process to establish VPAs is slow and will be of interest to major exporters to the EU market, the European Commission published in October 2008 a draft regulation aimed at recognizing the efforts of producers and traders that invest in ensuring the legality of their timber products. Operators placing timber and products made thereof for first time on the EU market will have to demonstrate due diligence in order to minimize the risk of allowing illegally harvested timber to enter the supply chain, be it from EU countries or elsewhere. It is expected that this type of regulation will be approved imposing significant new requirements on suppliers and importers in terms of provision of information, control systems, risk management, audits, and monitoring organizations.

In 2008, the United States of America amended the Lacey Act to combat illegal logging by making it unlawful to import, export, transport, sell, receive, acquire, or purchase any plants or products made from plants that were harvested or taken in violation of a domestic or foreign law. The Act gives the US government the power to fine and jail individuals and companies that import timber products harvested, transported or sold in violation of the laws of the country in which the timber was originally harvested. As regards trade, due diligence systems will have to be applied to provide assurance to importers that, when buying wood and wood-based products, they will not be prosecuted. The new legislative measures in the USA and the EU will have the potential to provide a robust incentive for tropical timber producers and exporters to stamp out illegal practices in forest management and timber trade and encourage them to make rapid progress towards demonstration of legal compliance.

The US and EU regulations are not identical and will be applied in parallel which may pose problems to exporters. There are several bodies which are presently applying their own standards or broad definitions of legality which is likely to create confusion among producing countries and can cast doubts on effectiveness of regulation. Common approaches or standards at international level could facilitate implementation but flexibility is needed to adapt them to local conditions which vary extensively in terms of the legal framework and institutional set-up. The goal for many countries is likely to be to carry out legality assurance through national systems rather than relying on private sector service providers. Common approaches can facilitate building up such national systems and thereby trade, provide an adequate framework for various private sector initiatives, and promote reliable communication on wood and wood-based products which are demonstrably legally produced.

4.6.3 Trade Measures in Combating Illegal Logging and Trade

Notwithstanding the potential of these approaches, we note that trade measures are only second best instruments for improving forest governance in exporting countries, as they do not address the underlying causes of illegal operations, which we will address in the following chapter of this book. Without tackling the main drivers through national action, trade related measures may have only a limited impact. In addition, the trade-related measures tend to have loopholes such as trade diversion, timber laundering and so on, which reduce their effectiveness. But, as we will see in Chapter 5, fundamentally it is often a question of lack of incentive and political will to control deforestation and forest degradation which creates the space for corruption and illegality, rather than it being the other way round – as it is frequently claimed to be. We have noted earlier that trade in forest products plays only a marginal role for many developing countries, while domestic markets are not generally sensitive to environmental concerns.

However, the situation is changing to some extent; for example, the Brazilian and Mexican governments have issued public procurement policies for timber which make reference to legality and sustainability indicating that domestic markets in developing countries are also being sensitized to the issue of illegal logging.

4.6.4 Can Certification Impact upon Illegal Logging?

Voluntary certification is a useful instrument for monitoring and verification of legal compliance of forest management and the chain of custody from the forest to the market. However, it was not *designed* for verifying legal compliance (in spite of the fact that all certification standards include legal compliance as their base line requirement) and therefore special verification systems have been developed for ensuring legality. There are, however, many hurdles to be addressed, not least the lack of a common approach to definition of what constitute legality and to minimum requirements of reliable legality assurance systems. In general, there will need to be a multilateral solution to the problem of trade regulation, rather than relying on unilateral importing country measures like the Lacey Act or bilateral arrangements like the EU voluntary partnership agreements. This would facilitate implementation by companies which supply different international markets and not put exporters into different categories depending on whether their country is participating in a bilateral arrangement or not. In addition, it should be ensured that these instruments do not become disguised measures of discrimination, a concern already raised about forest certification which is a de facto prerequisite for access to some markets.

The Convention on International Trade in Endangered Species (CITES) is targeted at protecting individual species against overexploitation due to export market demand. Its control procedures in the timber sector have proved to be problematic. In addition, CITES has several inherent limitations to solve this kind of problem on a broader scale such as few listed tree species, paper trail-based verification, high transaction costs, policy conflicts, institutional deficiencies, frequent corruption in tropical countries, etc. CITES as a measure of last resort to protect endangered species is a potentially increasingly powerful instrument in timber trade in the future as more species are included in its Appendices and electronic means make cooperation between customs authorities more effective. However, recent proposals to make forest certification mandatory under CITES would gradually lead to practical elimination of timber from natural tropical forests from international trade which would have a detrimental impact on forest communities and the survival of their forest resources the value of which would be drastically reduced in such situations.

4.6.5 Taking Stock

A level playing field for international trade in forest products could be built up through a combination of regulatory and voluntary measures. Monitoring technologies are rapidly evolving and other tools exist but trade measures are only a complementary element in solving the governance problems in exporting countries. They can however provide a strong signal for the need for change to policy makers. Due to structural complexities of addressing corruption from which all economic activities suffer – not only forestry – strengthening of governance is usually a long tedious process; one which will need to be underpinned by high level political commitment to the linkage of larger economic and social developments to forest outcomes. Trade measures can help in providing elements for the toolbox of disincentives such as limiting market access for illegal products, and incentives such as ensured access for legal products to public procurement markets.

4.7 New Opportunities and Challenges for Trade in the Valuation of Forests

Over the last decade interest has grown in regulatory, market-based and other voluntary mechanisms for payment for environmental services (PES) from forests. They are already a significant source of funding in many developed countries for conservation of watershed conservation and biodiversity but their greatest potential in developing countries will be in climate change mitigation and adaptation through increase or protection of carbon stocks in developing countries, and we will return to this subject in some depth in Chapter 7.

Globally, the actual impact of PES systems is presently still marginal both as a subject of trading or source of funding. The largest potential for market-based payments for environmental services is apparently related to maintenance of carbon pools through reduced emission from deforestation and forest degradation (REDD) and carbon sequestration in forests. The latter is currently possible through the Clean Development Mechanism (CDM) but by early 2009 only one project had been formally approved. The voluntary forest carbon market is still limited, estimated at US $50 million in 2008 (Hamilton et al. 2008) but it is growing.

With a few exceptions in Latin America (mainly Costa Rica, Mexico and the Andean countries), non-climate related PES mechanisms still play a limited role which is, however, growing. Various estimates have been presented on the potential size of the PES mechanisms to mobilize funding in developing countries (El Lakany et al. 2007; Bishop et al. 2008) but these estimates are highly speculative. The actual development of market-based PES mechanisms in developing countries has been slow for several reasons including inadequate policy and regulatory frameworks, weak market creation and promotion, limited engagement of suppliers (forest communities and landowners), a lack of technical and business management capacities, and so on (Bishop et al. op cit). Payment schemes may therefore have to rely on domestic public sector funding and international support in the short and medium terms but in the long run the prospects for market-based solutions appear bright and these could offer a significant potential measured in billions of dollars for sustained financing of forest environmental services.

Expansion of PES mechanisms could occur if schemes were able to demonstrate clear additionality (i.e. incremental conservation effects vis-à-vis predefined baselines), if PES recipients' livelihood dynamics are well understood and if trade-offs between conservation and income generation are balanced. PES mechanisms have both potential and risks as regards poverty. They may be best suited to scenarios of moderate opportunity costs on marginal lands and in settings with emerging, not-yet realized threats for forests. PES mechanisms are a win-win instrument as they can benefit both buyers and sellers while improving the natural resource management by internalizing sustainability costs. However, they are unlikely to fully replace other conservation instruments (Wunder 2007).

It is clear that PES mechanisms will be ineffective unless the legal, policy and institutional framework is improved, since lack of secure tenure, weak compliance, and corruption, increase risks and transaction costs. For this to happen, developing countries will need financial support for necessary upfront investments to install adequate legal and policy frameworks, to establish necessary institutional arrangements, to set up the transaction mechanism, to build capacity among actors (including forest owners and communities), and to raise awareness among stakeholders and the general public. PES mechanisms, though not a panacea, can help address the market failure problem of forestry and provide a critical element of revenue stream for SFM.

An effective and equitable solution to a public goods problem (e.g., ecosystem protection) will not be possible without appropriate compensation for the public good providers and effective regulation of the environmental and social externalities, and governments and the international community will need to play a much more effective role than they have done to date, to achieve this. Support is needed to generate realistic understanding of the possibilities of PES schemes, necessary preconditions for their effective implementation, and the needs for financing of upfront investments in capacity building, information system, setting up of appropriate voluntary and regulatory payment mechanisms with intended equity impacts. The importance of sovereignty issues in the context of developing a PES mechanism has been flagged in many intergovernmental negotiations on forests.

4.8 Conclusions

Trade is always driven by the motive of private gain. Trade liberalization alone does not therefore necessarily promote sustainable management of forests. However, through measures targeted at mitigation of possible negative impacts and enhancement of trade's potential positive impacts full compatibility can be established. This can be achieved by non-discriminatory market requirements which ensure that forest management for production of timber and non-timber products incorporates necessary provisions for maintenance of public goods such as biodiversity or climate change mitigation. However, in practice the situation is complex, due to the great diversity of local situations and the problem of free riders: the lower costs of producers who do not implement costly environmental and social standards to ensure sustainability of forest management but who benefit from market access. Both regulatory and voluntary measures can be applied within the framework of intergovernmental agreements on trade rules, forests and environment.

Regulation sets the minimum requirements for forest management. Voluntary instruments like certification of sustainable forest management should by definition build upon these requirements. Otherwise, the existence of multiple standards will tend to create uncertainty and sometimes even confusion among forest owners and managers. In order to expedite the change in forest management a number of countries have opted to internalize the certification standard requirements in their legislation. This blurs the general borderline between market-based and regulatory measures, but it advances the progress towards sustainability.

It is emphasized that sustainability of forest management is a moving target. It reflects the environmental, social and economic values given to forests in a specific country context at a given point of time. These values are dynamic as they change constantly which calls for periodic revision of what is understood as forest sustainability. Participatory processes in setting certification standards and their periodic adjustment are a useful tool to convert the emerging value changes into pragmatic guidance for forest owners and managers on what they should achieve.

The current emphasis on legality in market requirements may divert attention away from the concept of sustainability. This is a paradox as market-based instruments like forest certification have been introduced to correct policy and market failures. This is further associated with the failure of certification to deliver its promise as a measure to use markets to internalise the full costs of sustainable forest management in developing countries.

A common approach to contain illegally produced timber trade could be explored (e.g. global minimum requirements for national forest-related legislation and "harmonized" options for verification of legal compliance and legal origin). Past experience on international policy processes suggests that the case for international agreement on minimum requirements for national forest/environmental legislation could probably only be possible within an environmental context, rather than an economic context.

References

ATHENA Institute – FORINTEK (undated) Some Environmental Effects of Using Wood Compared to Other Materials. http://michigansaf.org/ForestInfo/MSUElibrary/RawMaterialEnergy.PDF. Accessed June 2009

Bishop J, Kapila S, Hicks F, Mitchell P, Vorhies F (2008) Building biodiversity business. Shell International Limited and IUCN, London, UK and Gland, Switzerland

Cashore B (2002) Legitimacy and the privatization of environmental governance: How non-state market-driven (NSMD) governance systems gain rule-making authority. Int J Pol Adm Inst 15(4):503–529

Chomitz KM, Gray D (2006) Roads, land use, and deforestation: A spatial model applied to belize. World Bank Econ Rev 10(3)

Contreras Hermosilla A, Dornbush R, Lodge M (2007) The economics of illegal logging and associated trade. Roundtable on Sustainable Development. OECD SG/SD/RT(2007)1/REV

El Lakany H, Jenkins M, Richards M (2007) Background Paper on Means of Implementation. Contribution by PROFOR to Discussions at UNFF-7, Apr 2007

FAO (2006) Responsible Management of Planted Forests. Voluntary Guidelines. FAO Forest Department. Planted Forests and Trees. Working Papers. FP/37/E. Rome

FAO (2008) Forests and Energy Key Issues. FAO Forestry Paper 154. Rome

Hamilton K, Sjardin M, Marcello T, Xu G (2008) Forging a Frontier. State of the Voluntary Carbon Markets 2008. Report by Ecosystem Marketplace & New Carbon Finance. 8 May 2008

ITTO (2007) Community-Based Forest Enterprises. International tropical Timber Organization Technical Series 28. Yokohama

ITTO/FAO (2005) Best Practices for Improving Forest Compliance in the Forest Sector. FAO Forestry Paper 145. Rome

Kaimowitz D (2003) Forest law enforcement and rural livelihoods. Int Forestry Rev 5(3):199–210

Katila M, Simula M (2005) Sustainability impact assessment of the WTO negotiations in the forest sector. European Commission, DG Trade

Li R, Buongiorno J, Turner JA, Zhu S, Prestemon J (2008) Long-term effects of eliminating illegal logging on the world forest industries, trade and inventory. Forest Policy Econ 10(2008): 480–490

Purbawiyatna A, Simula M (2008) Developing Forest Certification. Towards Increasing the Comparability and Acceptance of Forest Certification Systems Worldwide. *ITTO* Technical Series 29. International Tropical Timber Organization. Yokohama

Tacconi L, Boscolo M, Brack D (2003) National and international policies to control illegal forest activities. CIFOR, Bogor, Indonesia

UNECE/FAO (2008) Forest products annual market review. Geneva Timber and Forest Study Paper 23. Geneva

World Bank (2006) Strengthening Forest Law Enforcement and Governance. Report No. 26638-GLB. Washington, DC

Wunder S (2007) The efficiency of payments for environmental services in tropical conservation. Conserv Biol 21(1):48–58

Chapter 5
Deforestation: Causes and Symptoms

Abstract This chapter addresses a central tenet of this book: it is essential, when considering measures to address deforestation in developing countries, to separate the basic causes of deforestation from symptoms of it. This may seem basic, but there has been a strong tendency for some organizations and civil society groupings to view the problem from an ideological position or institutional preference, and misdirected effort has been the result.

The chapter reviews data and information on the main immediate causes of forest loss: agricultural technology and expansion; rapid growth of commercial plantations and grazing; mining; fuel collection and others. To a large extent, these are proximate causes of forest loss: they could be done in ways that are not destructive of forests, and the reasons this does not happen are more fundamental. This is why measures directed at proximate causes – such as boycotts in developed markets against agricultural products that have caused deforestation – are likely to simply displace one cause of deforestation for another. Illegal logging is more difficult to analyse: information on it is poor, it is highly variable in the manner in which it occurs, and it can be difficult to assign specific instances to particular interest groups.

More effort will be needed to identify the motives and incentives driving all groups having agency in a given deforesting situation, and then determining what will increase the value of sustainable alternatives for all involved.

In Chapter 3 of this book, we reviewed the scale and nature of forest loss globally, and in Chapter 4, we discussed the impact of trade on forests. Obviously, there is more at work than trade pressure in the large forests of the developing world, and we must now attempt an outline of the other influences acting upon these forests – especially tropical rainforests – in the countries where there are large concentrations of this remaining.

5.1 Rainforests: A Tragedy of the Commons?

In general terms, it is simple to explain the continuing loss of forests, in terms of ecologist Garrett Harding's concept of "the tragedy of the commons" (Hardin 1968). Hardin demonstrated its application to a wide range of activities which utilize natural resources (land, water, air), drawing on Aristotle, Hobbes and William Forster Lloyd, a political economist at Oxford University, who in 1832, when observing the devastation of commonly owned (as opposed to privately owned) pasture land in England, asked:

> Why are the cattle on a common so puny and stunted? Why is the common itself so bareworn, and cropped so differently from the adjoining inclosures?

Hardin reasoned that in such a case, where animals are grazing on an open access, common area, their individual owners will add to their flocks to increase personal wealth. Although the degradation for each additional animal is small relative to the gain in wealth for the owner, if all owners follow this pattern the commons will ultimately be destroyed. He concludes that each individual is engaged in a system that induces an attempt to increase the herd without limit.

Since Hardin published his article, the concept has been applied by resource economists and others to explanation of numerous environmental catastrophes, and even to other disasters: for example, some have attributed the savings-and-loan (S&L) crisis in the United States in the 1980s to the formation by the US Government of the Federal Savings and Loan Insurance Corporation (FSLIC). This effectively took on the liability of an S&L going bankrupt, by guaranteeing its depositors that, in that event, the FSLIC would repay their deposits, using taxpayers' funds. In effect, this turned taxpayers' funds into a commons, to be exploited: an incentive was created for managers of S&Ls to invest in high gain-high risk investments, and for depositors not to question such decisions. Some readers may detect similar elements in the explosion of sub-prime lending which triggered the financial meltdown in the United States, and the resulting global economic crisis which gathered momentum in 2008–2009: the commons in this case being the absence of effective oversight of risk in these markets.

In effect, Hardin's analysis was an application of the idea of the rational economic actor, or *homo economicus*, which has featured in economic literature since the days of Adam Smith. The dilemma, according to Hardin, was that if one member of a community dependent upon a *common pool resource* (forests, fisheries, oilfields, grazing lands, water resources and others are commonly used in this way) limits use of that resource – in order to conserve it and render it sustainable – but neighbouring users do not, then the resource collapses anyway, and the individual who limited use will simply miss out on the short term gain that others derive from overuse until the resource system fails in some way.

Elinor Ostrom, joint winner of the Nobel Prize for Economics in 2009, has, along with many others, challenged this notion (Ostrom, 1990). Her work has emphasized that humans can, and often do, interact with ecosystems to maintain long-term sustainable resource yields. She has examined conditions and rules operating in many situations in the field which in fact have led to successful and sustainable use of a common pool resource without resort to privatization or state ownership of the resource. Indeed, in a book edited by Ostrom and others (Dietz et al. 2002) many examples of successful management of a common pool resource are presented, and the editors argue that there are even evolutionary outcomes – in addition to socially determined ones – which suggest that group action can in certain cases lead to superior survival, compared to that derived from individual actions. The editors also point out that sometimes nationalizing a common pool resource – usually on grounds of creating a single owner who will act in the long term interest – actually makes things worse, especially when it involves outcomes such as rejection of local management institutions that have worked, and poor management and monitoring of the resource in the field by the responsible state agency. Often, a de facto open access problem is created by such measures.

It will be evident to readers that natural forests under the management of state agencies in many countries are a case in point. In Chapter 2, we noted the tendency of many national planners and their economic advisors to essentially "borrow from the future", in terms of sacrificing the future health and sustainability of some natural resource systems, in order to finance present economic growth. Global economic growth since the Industrial Revolution has been built on the development of effective markets for valuing and exchanging capital, human resources and other factors of production, but with highly ineffective markets – or no markets at all – for the natural resources that are fundamental to the maintenance of this economic system.

This continues in many countries – developed and developing – today. In Chapter 6, we will examine this issue by reference to interpretations of the Kuznets curve effect which suggest that under conditions of sustained economic growth, environmental degradation will eventually being to decrease. We will note there that if this process plays out at all in countries we are interested in, it will only do so after a long period of growth. In the meantime national policymakers may be effectively rendering natural resource systems as unsustainable open access commons, with predictable results.

The fate of the tropical rainforests can be seen as one of the largest examples – perhaps surpassed only by the world's oceans – of the de facto introduction of a tragedy of the commons, created as rainforest is managed as a state-owned resource. The instruments through which it has occurred include: a lack of adequate regulation, and enforcement of that regulation, leading to illegal extraction of logs; unsupervised (if theoretically legal) logging operations; officially sanctioned clearance of forested land for conversion to other purposes, without adequate analysis of the environmental and economic costs and benefits of doing so; and activities by other groups that are incompatible with retaining forest cover, in areas which are designated – in law and frequently even in national constitutions – as being intended for perpetual use and protection as forest. The very fact that it is not uncommon to find some or even all of these activities occurring simultaneously, or in a closely sequenced pattern, on a given piece of forest, will emphasize the nature of the problem.

The question is: what can be done about this?

A prior question that must be addressed, in advance of this primary one is: *what is causing deforestation, on any given site?* An important argument that we will make later in this chapter is that we must pay close attention to what is an underlying *cause* of deforestation, and what is a symptom. As we will see, this is not necessarily simple to make this distinction, and as we noted in Chapter 1 of this book, it tends to be the area of debate where many ideological preferences come into play. It is also, we will suggest, a matter which differs widely between different countries: a large export market based cash crop being grown in one country may be an underlying force of deforestation, whereas in another it may in fact be much more of a symptom, rather than a cause of deforestation.

First, we need to review the more common explanations of deforestation that have been advanced. As recently as a decade ago, deforestation was generally believed to be primarily a function of poverty: subsistence activities including shifting cultivation, gathering of plant material, charcoaling and other fuelwood usage. Much of the language which appears in the agreements on sustainable forestry discussed in Chapter 6 of this book focuses on these elements. These causes of forest loss are still present in many cases, and may be prevalent causes in some, in parts of Africa, and Asia, but particularly in the high deforestation countries, such as Indonesia and Brazil, other reasons for forest loss have become prominent.

5.2 Agricultural Technology and Deforestation

In a collection of studies edited by Angelsen and Kaimowitz (2001) a number of field studies and analyses were reviewed which collectively take a closer look was taken at one emerging factor: the linkages between agricultural technology and deforestation in the tropics. The broad findings from these case studies are presented (in highly summarized form) below:

- There tend to be trade-offs, rather than win–win outcomes, between forest conservation and technological progress in agriculture, in farming areas near forests.
- New agricultural technologies can stimulate deforestation when demand for the resulting output is not highly responsive to price – in other words, when prices are not depressed as supply increases: this tends to be the case when a significant proportion of the product is exported.
- New agricultural technology attracts migrants, and this will increase deforestation in nearby forested areas.
- Most farmers operating near the forest frontier will experience both capital and labour shortages. They will use labour-saving technologies to overcome this where possible so as to increase their output, and the result is likely to be deforestation in the nearby area.
- When agricultural expansion becomes capital intensive, technical change increases and can stimulate deforestation, if carried out in areas near forests.

On the other hand such developments in areas well-removed from forests can have the effect of reducing deforestation, as the centre of agricultural activity gravitates to these developments.
- Agricultural smallholders tend to employ a mix of production systems; technical progress with intensification may reduce deforestation initially, but the additional earnings made possible may stimulate expansion of agriculture into forested areas.

5.3 The Impact of Burgeoning Plantation and Grazing Commodities[1]

The impact of transition from small to larger scale agriculture can be seen in the two largest deforesting countries in the world, Brazil and Indonesia. In these countries, poverty and smallholder agriculture has come to be viewed as a minor factor, while the great bulk of forest loss is attributed to growing wealth. Later in this chapter, we will argue that some of the manifestations of this (reviewed below) might come to be regarded as symptoms of deforestation, rather than underlying causes. However, we still need to know what they are, and what would need to be done about them specifically in each case: obviously, we will need to be sure that whatever is done to reduce deforestation does remove these activities from – or at least prevents their further expansion of them on – the forests we seek to reserve for protection or sustainable management.

- In Brazil, the impact of poverty and smallholder agriculture is believed to be responsible for 20% of total deforestation, while the remainder (see Chomitz 2006) is attributed to newer wealth generating activities – grazing and soy bean production in that case.
- Accordingly, in these countries, drivers of deforestation being identified have been seen to have largely shifted from local forces to international markets and demand (Nepstad et al. 2006). Globalization of investment flows are seen to be allowing increased volumes of capital to feed into economic activities leading to deforestation.

---

[1] Much of the data and analysis on production and export trends for beef cattle grazing, soy, biofuels and palm oil come from background papers prepared for The Prince's Rainforests Project (discussed in Chapter 7 of this book), by a project staff member, Andreanne Grimard. We are grateful to her, and to the Prince's Rainforests Project managers, for kind permission to use this material here. We note that the general conclusions and recommendations that we draw from this material, later in this chapter and elsewhere in this book are our own, and not necessarily those of the Prince's Rainforests Project.

The outcome, under this interpretation, is that economic growth and changing consumption patterns, population growth, and rising commodity prices – especially on the international market – are now driving demand for forested land and resources[2]. This, combined with static agricultural yields and declines in soil fertility, will continue to raise the pressure on forests.

These developments have followed chains of events which differ between the two regions where deforestation has been high:

- In the Amazon, the chain of events generally follows a common pattern: land is partially cleared for timber, and cattle ranching usually follows. When soy farmers look for new land to expand production, their preferred choice is often cattle pasture land as it is cleared, and has lower value than land under agricultural production. This pushes the cattle ranchers to clear additional land – usually deeper into the rainforest – to resume activities.
- In Indonesia, developers have been establishing rapidly growing areas of oil palm plantations on forested land. Under existing official regulations governing use of the forest land, under the control of the Ministry of Forestry, only degraded forest is supposed to be converted for such plantations but, in recent years it appears that in fact forests which do not meet the definition of degraded forests have also been used. There has also been extensive allocation of forested land for conversion to pulpwood plantations, to serve the demand of Indonesia's large pulp and paper manufacturing sector. The official intention has been that natural forest logs obtained in the conversion process are to be utilized for pulpwood, until the plantations can take over this task. In fact, in some large cases it appears that the pace of use of natural forest logs as pulpwood furnish for the mills has significantly outpaced the establishment of viable and vigorously growing plantations, and in that sense has become a semi-permanent supplier.

5.3.1 Cattle Ranching (Brazil)

Beef is consumed as a source of protein, in its own right or as an ingredient in other foodstuffs. Because of the relatively high GHG emissions cost of growing beef, the Intergovernmental Panel on Climate Change has recommended that beef could be effectively replaced by other less GHG intensive meats, or by other vegetative sources of protein. However, it seems that this shift in consumption taste will be relatively slow in coming, and therefore to the extent that beef grazing is a significant activity in terms of deforestation, it will need to be addressed in more direct ways, so far as forests are concerned.

As can be seen from Table 5.1 below, Brazil is the major exporter of beef globally, and these exports have been growing rapidly.

---

[2] The rise of commodity prices alone cannot be singled out as causing deforestation. In fact, commodity prices were going down from the 1960s to 2002, while deforestation continued unabated.

Table 5.1 Global beef exports (in 1,000 mt) (United States Department of Agriculture. April 2008)

	2004	September 2008/2009
1. Brazil	1,610	2,200
2. Australia	1,369	1,360
3. India	492	800
4. Argentina	616	535
5. New Zealand	594	525
6. Canada	603	450

Russia is the largest purchaser of beef, followed by Japan and the European Union.

In Brazil, about 70% of the total area of rainforest which has been deforested is now occupied by beef grazing pasture (Vera-Diaz and Schwartzman 2005). One third of Brazilian beef exports come from the Amazon.

That proportion has been growing over the recent years due to lower land prices. In addition, more lucrative land uses in Southern Brazil, including soy and sugarcane (including for biofuels), have pushed cattle ranching into the Amazon.

Cattle ranching in Brazil is a relatively low intensity land use, not rising much above one beast per hectare. It is estimated to have emitted 9–12 billion tons of GHG over the last decade, including land use change and enteric fermentation of the herd, but without taking processing and transport into account (Friends of the Earth 2008).

Infrastructure development in the Amazon region, including slaughterhouses and transportation links, has further encouraged the expansion of cattle ranching in the region.

Increasing demand and devaluation of the Real has led to a boom in beef production and exports over the last decade. In 2007, revenues derived from Brazilian beef exports reached $4.418 billion – a 13% increase on the previous year (Association of Brazilian Beef Exporters (ABIEC)), but it should be noted that the Real has revalued significantly against the US dollar since then. United States meat exporters forecast that Brazil will increase exports by 27 percent, totalling 3.05 million metric tons by 2017 (USMEF 2007).

5.3.2 Soy

Soy is used for a wide variety of purposes including oil, protein rich food (e.g. tofu), and as ingredient in feed for livestock. It can also be used for biofuels. It has remarkable market penetration into the international foodstuffs sector: over 60% of all processed food in Britain today contains soya. Breakfast cereals, cereal bars and biscuits, cheeses, cakes, dairy desserts, gravies, noodles, pastries, soups, sausage casings, sauces and sandwich spreads contain soya. It appears on food labels as soya flour, hydrolysed vegetable protein, soy protein isolate, protein concentrate, textured vegetable protein, vegetable oil (simple, fully, or partially hydrogenated), plant sterols, or emulsifier lecithin (Lawrence 2006).

Brazil is the second largest producer of soybeans, after the United States, which is expected to produce about 85.0 million metric tons of soy in 2008/2009. At the time of writing, Brazil was expected to produce about 62.5 million metric tons of soy in 2008/2009, of which around 27.5 million metric tons will be exported (principally to China and the EU). China's appetite for this product is enormous: it currently imports more than twice the soy imported by all G8 countries combined – despite being a soy producer itself. Chinese soy imports have grown 30-fold between 1995 and 2006: much of it is used in animal feed for beef, cattle, and pork.

Production of soybeans in Brazil has increased rapidly, due to: significant growth of international demand for soybeans; decreased production of soy in the US in favour of corn due to increased subsidies for biofuels (discussed in the following sub-section below); sharp increases in global prices; devaluation of the Brazilian Real; improvements in Amazonian infrastructure; and the development of soybean varieties suited to the Amazonian climate (Vera-Diaz and Schwarzman op cit).

The latter two developments in this list have contributed to an anuual increase of 15% of soybean production located within the closed-canopy forest region of the Amazon, from years 1999 to 2004 (Nepstad et al. op cit). Soy is mostly grown in flat and relatively dry zones at the eastern and southern margins of the Brazilian Amazon and is now spreading to the western and northeastern parts of the country.

It is important to note that by occupying pasture areas, soybeans displace cattle ranching from the margins further into the core of the Amazon region. Because of the soy boom, prices of land in Mato Grosso have quadrupled between 1999 and 2004, leading to land speculation and cattle ranchers moving into cheaper and more distant lands (Volpi 2007).

5.3.3 *Biofuels*

Biofuel is defined as solid, liquid or gaseous fuel obtained from biological material. Various plants and plant-derived materials are used for biofuel manufacturing. It can also be produced from gases from landfill, and recycled vegetable oil, but the main expansion in recent years has been from crops grown principally for the purpose of biofuel production; these are sometimes referred to as agrofuels, and they are classified as *first generation biofuels*.

There are two common strategies of producing liquid and gaseous agrofuels. One is to grow crops high in sugar (sugar cane, sugar beet, and sweet sorghum) or starch (corn/maize), and then use yeast fermentation to produce ethyl alcohol (ethanol). The second is to grow plants that contain high amounts of vegetable oil, such as oil

5.3 The Impact of Burgeoning Plantation and Grazing Commodities

palm, soybean, algae, jatropha, or *Pongamia pinnata*. When these oils are heated, their viscosity is reduced, and they can be burned directly in a diesel engine, or they can be chemically processed to produce fuels such as biodiesel.

Globally, biofuels are most commonly used to power vehicles, heat homes, and for cooking. Biofuel industries are expanding in Europe, Asia and the Americas. In recent years, the subject of biofuels has become controversial, for several reasons: First, they have been promoted for geopolitical reasons, such as reducing energy dependence on Middle Eastern oil supplies. Second, recent sharp rises in global food prices have been blamed on expanded first generation biofuel production taking up agricultural areas previously utilized for food production, although estimates of this impact on vary widely: the Food and Agriculture Organization suggests that 10% of food price increases can be attributed to biofuels; while the International Food Policy Research Institute claims it accounts for 30% of the increase. Thirdly – and of specific interest in this book – these fuel crops have also been criticized when expanding into forests and other environmentally sensitive areas. While biofuels have been promoted on the grounds of their ability to reduce the transport sector's net GHG emissions, when their production in the tropics involves land conversion of forest – especially rainforest – it will almost certainly produce higher GHG emissions than the alternatives.

However, the extent to which this will occur in the future, and the dynamics which will create this impact, are difficult to predict, because the linkages between biofuels and tropical deforestation are often indirect and thus very difficult to quantify. In Brazil, for example, sugarcane is produced in Southern Brazil, but as more land in the South is used for biofuels production, soy cultivation and cattle ranching are migrating to the Amazon frontier. In Indonesia, the critical issue, as we will examine further in Chapter 7 of this book, is the extent to which oil palm production can in future be located on non-forested land. At present, palm oil for biofuel production in Indonesia is only a very minor part of this question: about 5% of global palm oil is currently used as a biofuel. The cost of palm oil has become prohibitive for biofuel companies (US $1,300 per ton against a current break-even point for use of palm oil as fuel of US $800 per ton)[3].

Partly in response to the issues outlined above, there is now considerable interest in *second generation biofuels*, which are derived from a greater range of non-food feedstocks, and which feature higher yields and less stress on soil and water. As examples: wood and its byproducts can be converted into biofuels such as woodgas, methanol or ethanol fuel, and some agricultural residues, grasses, and even microscopic algae cultivated in specially constructed environments are possibilities. At this stage, these processes are not yet viable economically, and their environmental implications have also not been fully explored.

Less than 10% of biofuels are currently internationally traded (Thow and Warhurst 2007). Currently Brazil and the US produce 90% of global ethanol, and they are also

---

[3] Conversation with Mursalin New, Mission Biofuels 2008.

Table 5.2 Global palm oil production and exports ('000 mt)

	Global palm oil production ('000 mt)		Palm oil exports ('000 mt)	
	2004/2005	2008/2009	2004/2005	2008/2009
1. Indonesia	13,560	19,700	9,631	14,800
2. Malaysia	15,194	17,700	12,684	14,360
3. Thailand	820	1,400	81	500
4. Colombia	647	830	327	405
5. Nigeria	790	820	333	300
6. Other	2,509	2,752	1,497	1,530

large consumers of the product. Demand and supply of biofuels are largely driven by regulation, including targets and subsidies. Support for biofuels takes many forms including: market price support, excise tax exemptions, output payments, assistance to value adding factors, assistance to intermediate inputs, R&D grants, and support for consumption (see Table 5.2 above). The most important of the latter are excise-tax exemptions. The future of biofuels will largely be dictated by developments in these regulatory measures. A large part of their raison d'etre will depend upon how well they are able to genuinely reduce GHG emissions, without putting undue pressure on food prices.

Given this, it is difficult to predict the future of biofuels. It is generally expected that as oil prices rise biofuels will become an increasingly attractive alternative, and demand will rise further. Under current policies, the OECD projects areas for biofuel crops to increase by 242% between 2005 and 2030 (OECD 2008). Some observers foresee even greater expansion of investments in the biofuels sector: a Time Magazine article reported that investments grew from US $5 billion in 1995 to US $38 billion in 2005 and are expected to top $100 billion by 2010 (Grunwald 2008). On the other hand, NEF has reported a decrease of venture capital and private equity into biofuels from 2006 to 2007 (New Energy Finance 2008).

5.3.4 Palm Oil

Oil palm trees are native to West Africa, but are now grown throughout the tropics to produce large quantities of vegetable oil, much of which is traded through international commodity markets. Current demand for palm oil is predominantly for food products and cosmetics. Palm oil can be found in 1 in 10 products in UK supermarkets. It has increasingly replaced other sources of fats due to increasing concern with trans-fats. It is often labelled as vegetable oil. India and China's growing appetite for palm oil as cooking oil also plays an important role in the increase in palm oil production. In 2004, China imported 20.2 million tons and India 16.5 million tons, most of it used as cooking oil (see FAOSTAT data cited in Koh and Wilcove 2008).

China is by far the largest importer of palm oil, taking in 6,200 mt in 2008/2009, followed by India (4,700 mt) and the EU (3,950 mt), with Pakistan, Bangladesh and the United States taking a total of 4,530 mt in that year.

The overwhelming majority of palm oil production now takes place in Indonesia and Malaysia (approx 90%). Other producing countries include Nigeria, Thailand, Colombia, Cote d'Ivoire and Ecuador. Expansion of cultivation has recently been taking place in Brazil, Liberia and DRC. In West Africa, palm oil has traditionally been mainly produced by smallholders for local processing and consumption. However, some palm oil multinationals have showed renewed interest in the region. A Chinese company is reported to have acquired 3 million hectares in Democratic Republic of Congo and Sime Darby has a palm oil concession in Liberia. Unilever has large grown oil palm plantations in Ghana and Nigeria.

Great concern has been expressed by some observers about the impacts of oil palm plantation expansion on tropical rainforests in South East Asia: Koh and Wilcove (2008) have suggested that in Indonesia and Malaysia, about half of palm oil expansion occurs at the expense of rainforests. As will be seen later in this chapter, we will question the nature of causality between oil palm plantation and deforestation; however, there is no doubt that for a variety of reasons large areas of oil palm plantation have ended up on previously rainforested sites. There is equally little doubt that the formidable economics of palm oil in South East Asia, and probably elsewhere as well, will continue to drive rapid expansion rates for this crop:

- In 2004, Indonesia revenues from palm oil were around 1.7% of GDP (export value of $4.1 billion), and 5.6% of Malaysian GDP ($6.3 billion) (Koh and Wilcove 2007 op cit) As both prices and demand have recently gone up this number could be expected to rise.
- In Indonesia alone, approximately five million people derive their livelihood from palm oil.
- Prices for palm oil are said to have quadrupled over the last four years, due to rising demand. A significant part of this rise is often assigned to expected increases in biofuel production, but as pointed out in the **Biofuels** discussion earlier in this chapter, only about 5% of palm oil is currently used in biofuels.
- Oil palm plantations have yielded a net profit of up to US $6,300 per hectare per annum. This would seem to put opportunity costs considerably higher than estimates made by the Stern Review (Stern 2006), but note also the discussion of the oil palm example in Chapter 7.
- Global palm oil consumption increased by 6.6% to 35.3 mt in 2006. This trend continued in 2007 with an estimated 5.5% consumption rise (USDA 2006).
- In 2007, approximately 13% of palm oil exports went directly to Europe but a large share also goes to China and India and is processed in products that they export to Europe (VROM 2008).

5.3.5 *Forest Industry Plantations and the Pulp and Paper Sector*

Both Brazil and Indonesia have established large areas of plantations of trees, primarily for use in their pulp and paper manufacturing sectors, but also for use in the solid wood products industries. There are also large areas of commercial plantations established in Southern Africa.

Under the right circumstances, tree plantations can be beneficial for natural forests, by providing an alternative wood supply which can remove pressure on them. In the case of Brazil, about 6 million hectares of plantings have been established, most of it on open land, and little on previously rainforested sites. The majority of the plantings are of fast growing eucalyptus species, which provide pulpwood for Brazil's large pulp and paper sector, which is supplied almost entirely from plantation stock, rather than natural forest pulpwood.

Indonesia presents a very different situation. In this country, the theoretical attractiveness of plantations can seem potent: land is in short supply and is heavily contested in Indonesia, and establishing raw material resource systems which can produce log volumes at ten or twenty times the rate of a regenerating natural forest seems like a useful way to economize on the demand for forested land from the large and rapidly growing pulp and paper sector, and from other wood products sectors that can utilize plantation timber.

The Government of Indonesia began allocating significant areas of what is classified as Production Forest land as industrial plantation sites in the 1980s. Of the 23 million hectares of forest land allocated to all conversion purposes in Indonesia, as at 2006 about 10 million hectares had been allocated to industrial plantations, most of it (6 million hectares) for use as fast-growing pulpwood plantations, to serve the needs of Indonesia's rapidly expanding pulp and paper manufacturing sector; and the remaining 4 million hectares for longer rotation plantations to serve the needs of the Indonesian wood products sector[4].

However, only about 2.8 million hectares (a relatively small fraction of the total area allocated to the plantation programme) has actually been planted: most to pulpwood species. Even significant parts of this area may not be effective, growing plantation any longer. Although the scheme has existed for more than two decades – technically sufficient time for three rotations of pulpwood planting to have matured to harvest – the pulp and paper sector still only receives a minor proportion of its pulp furnish from plantation timber.

Many reasons are offered for the relatively poor outcome of the plantation program to date – unresolved land use conflicts in much of the forest area where plantations were located; inadequate technical application in choice of species, establishment and maintenance shortcomings; complications over land security introduced with decentralization in the late 1990s; competition for the same land from oil palm and other sectors in certain locations; and so on. It may also be the case that some sites selected for plantation were relatively unsuited to this purpose due to soil, topographic and geographical factors related to the location of processing and port facilities.

Some observers have suggested that there may also be fundamental constraints which investment in plantations shares with broader rural development in Indonesia:

---

[4] The source of this information is the Ministry of Forests, Indonesia (various documents).

social and cultural issues affecting the availability of land; economic policy and incentive issues which affect the availability of credit for this type of investment; and technical limitations on physical infrastructure, on the provision of capacity building in the plantation business, and on the viability of plantation establishment on barren or degraded lands, which are the most available.

There have been trenchant criticisms of the fundamental economic viability of the tree plantation sector in Indonesia. Maturana (2005) has analysed the economic costs and benefits of five large pulp plantation projects in Sumatra, and has concluded that, on the basis of an "efficiency price" of $40 per cum for pulpwood, the economic benefits of producing pulpwood is well below the costs of conversion of this land. She suggests that the allocation of 1.4 million hectares of forest land for conversion into tree plantations has generated losses of $3 billion for Indonesia.

Underlying all these explanations, however, is the fact there was a fundamental design flaw in the official plantation program – There was basically no effective price incentive applied. Concessionaires and industries who received licences to establish plantations were permitted to log large volumes of remaining natural timber from the allocated sites (which in the main have been many times larger in area than the intended area of plantation) at extremely low stumpage charges. Under such circumstances, it is rational for pulpwood users to maximize the volume of natural timber they use for pulp furnish, and to endeavour to attain rights over as much of this resource as possible for as long as possible, rather than to seek to replace it with plantation material as soon as possible. The situation appears to be that the plantation scheme has been, and remains, of more value to the recipient of a licence as a cheap source of natural forest raw material, than as a site for establishment of plantations.

Whether originally intended or not, this would appear to be the result of providing much larger areas of forest land than are required for plantation purposes, under the plantation licences. It is reasonable to suggest that large, financially viable companies would be capable of improving the efficiency of their investments in plantations, if the incentives to do so existed. Indeed, there are a number of examples of companies which have managed to establish and maintain good quality, high productivity plantations in Indonesia.

None of these observations are new in Indonesia. Some years ago Kartodihardjo and Supriono (2000) argued that timber plantation development policies in Indonesia legitimize the degradation of natural forests; that subsidies are ultimately unnecessary for the development of timber plantations; that tree crops plantation developers request – and receive – more land than they need, for the purpose of generating additional profits from the timber on lands to be cleared; that overlapping and chaotic forest land use classification systems work to the benefit of private plantation developers at the expense of rights and livelihoods of forest-dwelling people; and that government failure to recognize the rights of such people exacerbates this problem.

The ongoing consequence of this situation is that, after all this time, Indonesia's pulp and paper manufacturers are still reliant primarily on natural tropical forests for the expansion of their supply of pulpwood: past government policies and weak enforcement have allowed pulp mills to expand their processing capacity in advance of creating a sustainable plantation base.

The area of Indonesia's tree plantations is likely to expand significantly over the next years. As the availability of timber from natural forests is declining – especially in areas within commercial haulage distance of the large mills – pulp producers, plywood producers, and furniture manufacturers will be forced to turn increasingly to fast growing tree plantations as a source of raw material. There are signs that national pulp production capacity will continue to expand, requiring a further increase in plantation area. Over recent years, there has also been a growth in chipping mills that produce pulpwood chips for export, creating further demand for plantations.

In response to anticipated timber demand, the Ministry of Forestry has set ambitious targets for increasing the area of tree plantations, and has allocated over 10 million hectares for industrial plantation concessions, most of which are still undeveloped. In addition, the Ministry of Forestry has plans to establish another 5.4 million hectares of community based timber plantations up to 2016. Yet, the Government has publicly committed to a ban on the use of remaining naturally forested land for purposes of pulpwood supply, from 2009 onwards.

We have spent some time on this particular case because it is an example of where the outcome of a policy – increased pressure on the natural forest resource – has been the opposite of its stated intent – to produce adequate wood supplies from plantations to allow remaining natural forests to be managed sustainably. Not only has this led to extensive forest loss in the past, but the legacy remains in that large pulp and paper manufacturers will require ongoing supplies of pulpwood from somewhere, and although there appears to have been some increase in pulpwood supply from existing plantations in recent years, it seems unlikely this will close the supply gap created by a ban on further use of natural forest timber for this purpose. Some companies will be better positioned to move relatively quickly to plantation based pulpwood supply than others.

5.4 Some Other Factors in Deforestation

5.4.1 Mining

The extraction of valuable minerals found in rainforest areas can result in local deforestation. Although the area needed for mining can be small, it often leads to a domino effect. Mining operations generally entail the building of roads and other infrastructure. The inward migration of workers often leads to small scale deforestation as a result of building new homes, creating gardens to grow food and the extraction of fuel wood.

Many rainforest nations have important mineral deposits which have attracted national and foreign companies to develop the required infrastructure. The Democratic Republic of Congo features vast deposits of cobalt, coltan, copper, diamonds, gold, manganese, zinc. In Madagascar there are a variety of valuable minerals including gold and ilmenite. Minerals known to exist in the Amazon Basin include diamonds, bauxite (aluminium ore), manganese, iron, tin, copper, lead and gold. In Indonesia, mining of coal, gold, copper and bauxite has placed pressure on forested areas.

5.4.2 Wood Fuel

While many people in the developed industrialized countries take clean and abundant energy for granted, in many developing countries people still rely on wood to meet their basic energy needs. However, fuelwood is generally regarded as a weak overall driver of deforestation: most wood fuel is derived from trees grown for this purpose and litter from farm trees, or timber taken from agricultural clearing.

There are exceptions, however. Wood harvested for charcoal production and sold into urban markets can lead to deforestation close to urban areas or other centres with high population density. In Brazil, use of wood in charcoal which is manufactured for local iron-smelting activities is regarded as a serious cause of forest loss. In parts of Africa where alternative energy is not available or affordable, wood fuel over- harvesting can cause deforestation. For example, in Madagascar, deforestation has taken place not because of commercial logging, but because impoverished local people have no alternative fuel source. Only about 10% of Madagascar's original forest cover (about 6 million hectares) remains, and this continues to be cleared for wood fuel at a rate of about 200,000 ha a year.

5.4.3 Pioneer Shifting Cultivation

People who have neither the money nor the political power to acquire permanent holdings on productive lands, often settle along roads constructed in the rainforest for regional development and extractive industries. These people use the trees for building materials and slash-and-burn techniques to clear forest for short-term agriculture, planting crops like bananas, palms, manioc, maize, or rice. Denuded of trees, the productivity of the soil declines after a year or 2, and farmers move on to clear additional forest for more agricultural land.

There are many complex reasons why people move to the frontiers of the rainforests but the common factor appears to be lack of economic alternatives. Sometimes migration is encouraged by governments for economic and political reasons. Poverty and landlessness can lead people to seek a livelihood in rainforest areas. Population growth is often behind why people move to forest frontiers. For example, in the 1980s, Indonesia implemented a transmigration programme to alleviate some of the population pressures in the central islands, such as Java, by moving people to the outer islands of the country (including Papua, Sulawesi, Kalimantan, Sumatra). This resulted in increased population pressure in once sparsely populated areas.

As noted at the outset of this chapter, the international forests constituency once regarded shifting cultivation as one of the major causes of deforestation. However, more recently the practice has come to be seen as less damaging than originally thought (although in cases where population change is rapid in an already intensive shifting cultivation area, the cycling period can become foreshortened to the extent where severe damage to forests is done). In Africa particularly, some initial statistics and information about forest loss due to this cause have been revisited, and conclusions about resulting forest loss have been modified significantly.

5.4.4 Infrastructure

In order to improve the livelihoods of their people, the governments of rainforest nations instigate major infrastructure projects such as the building of roads and dams. If not planned carefully, this can often result in irreversible damage to the areas of rainforest in which they are located.

For example, the construction of new roads can involve the felling of forested land, but they also facilitate the movement of workers and materials involved in activities such as logging, agriculture and mining.

Plans to build dams in river basins in rainforest areas threaten forests with flooding. Currently, the biggest of these planned projects in the Amazon is the Tocantins River Basin hydroelectric project stretching over a distance of 1,200 miles. Significant areas of forest in Laos have been inundated due to construction of dams in the mountainous Eastern forested areas of that country.

5.5 Illegal Logging

Of all the causes of forest degradation and deforestation, in the tropics and elsewhere, illegal logging is the most ubiquitous. Sometimes, we suggest, illegal logging will occur purely for its own sake, since it is a lucrative and in many cases a relatively low risk activity. At other times it will accompany the other conventional drivers of deforestation (as we have termed them) that we have examined above, in which case illegal logging becomes the means by which political or social forces which produce deforestation are manifested. It is also the most difficult activity to quantify, virtually by definition: the very nature of illegal logging is that it operates outside the laws and regulations that govern forest activities, and therefore outside the means by which legal operations are assessed and measured.

As if this were not enough, there are also major differences in how different groups define illegal logging. Some will suggest that any logging operation which breaches

5.5 Illegal Logging

any regulations applying silvicultural and logging rules of practice are by definition illegal, and should therefore be closed down. Others will argue that the term illegal logging should only apply to operations which are completely unsanctioned by law: possessing no official permit or licence to log in the area in which they are found to be doing so. Obviously, there are many variations on these themes, and many shades of meaning within them.

While the motives for expansion into forested areas of beef grazing in Brazil, or of oil palm plantation in Indonesia can be clearly assigned to a particular interest group, illegal logging often cannot. In some cases, significant illegal logging occurs on the periphery of legal logging operations, either directly by the logging firm itself, or by agents with commercial incentives provided by that firm, who can access the forest via transport infrastructure already in place. In others, illegal logging occurs as a result of the activities of operatives hired by large commercial interests who may have no direct activities in the area in question at all; processing firms, or traders who intend to smuggle the logs out of the country, may be involved. In virtually any situation where rainforests are reasonably accessible to significant numbers of local populations, illegal logging for basically livelihood purposes – either as part of a shifting cultivation process or not – will be present. Often, this "background" level of illegal operation is regarded by official agencies and/or by civil society groupings as inevitable, and tolerable to an extent given the poverty conditions that usually apply.

Clearly, whatever the origin of illegal logging, official forest management agencies have an incentive to minimize figures released on the level of such operations within their area of responsibility, since this goes directly to the competence and probity of the agency itself. Meanwhile, environmental and social groups who perceive official state forest programs as unsustainable and bad for local communities will choose illegal logging estimates from the high end of the interpretation scale, to emphasize the urgency and gravity of the problem.

After consulting many of the published and unpublished sources of data and analysis on the illegal logging problem, we confess ourselves unable to select from the range of figures encountered an estimate of what the actual level of forest loss directly from this source might be. Nor can we determine, at this point, whether the many new government and donor initiatives which have been launched to address this problem (see discussion of the FLEGT programmes in Chapter 7 of this book) have had a significant impact upon it.

We do, however, share the general feeling reflected in the international forests constituency that the illegal logging problem is serious, on the basis of our view expressed earlier that ongoing deforestation must in most cases involve illegal logging, either on its own or accompanying some other dynamic of deforestation. The high levels of deforestation that we have noted prevail in the tropical rainforests must therefore indicate significant levels of illegal logging, since such levels of deforestation are usually not intended in official forest programmes (although we should note that in some cases, directly or by implication, some level of deforestation is part of the official plan).

Below, we list some of the assertions that have been made about the illegal logging problem by various international agencies and interest groups:

- In estimates made in 2002 in support of its revised global forests strategy, the World Bank suggested governments lose about US $5 billion in revenues annually as a result of illegal logging while overall losses to the national economies of timber-producing countries add up to an additional US $10 billion per year. A later estimate by the Bank puts the value of *illegal* timber at US $7 billion annually in lost assets, lost revenues and unpaid taxes.
- Approximately 40% of trade in illegal timber is attributable to G8 country imports. Japan and the USA are responsible respectively for a half and one quarter of those G8 imports. Much of this illegal timber is imported as logs (mainly Japan), sawnwood, plywood, furniture, and flooring (see the Globaltimber (undated) site for information on this).
- According to the UNDP and DFID an estimated 73–88% of all timber logged in Indonesia is illegal. Less than 20% of this is smuggled out as logs, and the remaining wood is processed in saw, paper or pulp mills and later exported. These mills have a capacity of two to five times greater than the legal supply of timber.
- Some sources estimate that as much as 500,000 ha (42% of annual logged area) of forest in Indonesia are illegally logged each year at a loss as high as US $3.5 billion in revenues to the government (see for example MongaBay 2009). The logging process often results in the damage of almost twice the volume of the harvested trees.
- China buys half of all internationally-traded tropical logs, for a total value of $8 billion (DFID 2007). One out of every two is illegally harvested. Around two-thirds of the imports come from Asia and the Pacific – with Malaysia, Papua New Guinea, Myanmar and Indonesia being major suppliers – but China is also having a big influence on the timber trade in Africa and Latin America (DFID 2004).

5.6 Commentary on Some Corrective Options

A number of approaches, based around the drivers discussed above, are commonly encountered in the literature on deforestation in the tropics. All of these, we suggest, could potentially have some effect on reduction of deforestation, but only if they are applied within a coordinated strategy, rather than on an ad-hoc, ideology-driven basis.

5.6.1 International Demand Management

As discussed in Chapter 4, it is commonly noted that there is presently little awareness at the consumer level of the damaging impacts of production of some of these products on rainforests in the large international markets for products such as beef,

5.6 Commentary on Some Corrective Options

soy, palm oil and other products. As a result, there is little incentive for retailers to seek sustainable sources of these products; in many cases, it is noted, consumers may not even be aware of the presence of these items in products they purchase.

Not surprisingly, this means that the idea of marketing products which originate from sustainably managed environments at a premium price has not taken hold in many cases.

The response from some groups concerned about this is sometimes to promote awareness campaigns, or even to press governments in importing countries to regulate entry of the offending products, in an attempt to force suppliers to alter their production practices. In our view, attempting to manipulate demand in large consumer nations for these products as a stand-alone initial response is an extremely blunt instrument, likely in most cases to fail. While it may appear to activists in consumer countries as a benign and positive measure, it is much more likely to be seen in the supplier countries as an attempt by developed nations to pass on the costs of their emissions reductions and other global environmental misdemeanours onto developing nations.

If in fact the demand measure being promoted is *all* that is being done, then it is difficult not to agree. Containing the leakage to consumer markets with less concern in this matter – including some large developing country markets (which raises the issue of who bears primary responsibility for GHG reductions and other measures) will be problematical.

Awareness campaigns may have some effect at the individual retail firm level, just as certification of timber products can have some impact in the same way. However, the further any measure of this nature moves away from having an impact on the consumer market at an aggregate level, the less likely it is to have any impact at all, beyond shifting product around between the various international buyers, rather than exercising any impact on supply production methods. In this respect, consumers should recognize that a good outcome from the viewpoint of an individual retailer who has chosen to pursue the certification path does not necessarily imply a similar result for the market as a whole, or any subsequent significant impact on aggregate production in the supplier countries.

Combined with proactive and practical measures in the supplier countries to re-orient production systems to more sustainable models, demand measures in consumer countries could have more impact, but they should certainly follow such measures, and preferably be done in collaboration with the supplier country, rather than take the leading role.

We suggest a thought experiment here: if there is a strong conviction in a given consumer country that that a given food product is causing widespread deforestation in supplier countries, and a strong desire to address this problem, then what approach would be likely to work best? Exclusion via certification or other measures of that product from the market? Or an offer to the supplier countries to cover the cost of transition to sustainable production of the product in question, perhaps by means of a levy raised on all foodstuffs in the consumer country?

There is nothing inherently wrong with the principle of using market forces to foster sustainability, but it is an approach which is unlikely to succeed unless the burden of cost of moving to more sustainable forms of production is at least shared, if not taken on completely by the consumer country, rather than being thrust upon the supplier country. Attempting to force such a country into compliance in this way will run the risk of producing undesirable impacts: First, the government and people of the producer country will in general see this form of sanction negatively, lowering the chances of cooperative action. Second, the country will seek other markets, less discriminating, for its products; these markets may be less lucrative, which will drive the producer county to lower costs even further, possibly through even more environmentally damaging activities[5]. Eventually, this becomes (in the eyes of the producer country) a self-solving problem; forests will continue to be lost, ultimately to the point where other crops are produced on the cleared land. Sooner or later, consumer countries will buy those products; it will be unrealistic (if not impossible) to keep tabs on all land originally subjected to deforestation by the original product involved, and exclude importation of anything produced on that land forever.

For reasons such as this, in our view, some of the tendencies and proposals of groups who have advocated the trade approach have in fact drifted from being an idea, more into the realm of being an ideology: it confuses cause-and-effect in this area, and it runs strong risks of being counterproductive, unless designed and implemented in consultation and cooperation with producer nations, with adequate back-up and support measures to ensure that the right incentives are created.

5.6.2 Eliminating Perverse Production Incentives

A more appealing approach would be for countries which are producing these products in an unsustainable way to eliminate fiscal and regulatory measures that encourage inefficient and destructive production practices. There are many such counter-productive measures in place in the forests and agricultural sectors in the developing countries. In some cases, tenure arrangements which prevent local communities from participating in sustainable production systems and drive them to unsustainable and illegal activities for basic livelihood are a major problem. Changing actual tenure to entitle such communities is a desirable long term goal, but measures to improve their rights of access and usage of forested and other lands – and in some cases to actively limit the access and usage rights of other interest groups in the near term would need to be implemented much more quickly, to have any impact on deforestation. However, experience shows that reform processes involving actual changes in formal tenure are politically sensitive, technically demanding and costly and therefore their implementation has generally been extremely slow.

---

[5] Ultimately, this will not be a long term solution, as the standards of global trade tend to harmonize over time – but significant damage may be done in the meantime.

Many agriculture sectors in developing countries receive subsidies and guaranteed prices for output that lead to overproduction of these products – often at the expense of forested land. Sometimes international donor aid projects have in fact exacerbated these tendencies, and financed further expansion. The taxation system is often used in the same way: for example, until relatively recently in Brazil, taxes on unused forested land held by private owners were higher that for cleared land (Volpi 2007 op cit) with obvious results in terms of forest clearing. Inconsistent or uneven application of land use regulations and enforcement measures can also accelerate forest degradation, either as illegal logging operations are poorly controlled, or agricultural expansion is tolerated by local and national authorities.

Obviously enough, improvement in both the design and the implementation of policies around these issues would result in greater survival of forested areas. However, as discussed in Chapter 2 of this book, we need to exercise great caution when attempting to interpret why such improvements have not occurred – or are painfully slow in arriving – in many rainforest countries. Until there is an alignment between local and national perceptions of the value of intact forest on the one hand, and how this is perceived at international level, there is little likelihood of change. We will discuss this issue further in Chapter 7.

5.7 Separating Causes and Symptoms

Although we have discussed factors which have been associated with deforestation and forest degradation individually in this chapter, we do not want to encourage the idea that these activities operate separately, or that they might be best dealt with by separate corrective actions. As noted at the outset of this chapter, the specific activities which lead to deforestation can operate in groups, simultaneously, or in close sequence. As we will discuss further below, elimination of a given driver of deforestation could conceivably have no impact on deforestation at all: it could result in replacement of that particular cause by another, or it could simply be that the sequence of activities creating forest degradation and deforestation continues unabated.

We are by no means the first to recognize that the reasons and forces behind deforestation are related to complex and overlapping competition for land and resources. Geist and Lambin (2002) have classified drivers of deforestation in the tropics into two basic categories: proximate causes and underlying driving forces. The former impact directly on the forests: infrastructure, agricultural extension, wood extraction and so on. Underlying causes are demographic trends, economic decisions, technological, policy and cultural factors.

Their compilation is shown in Fig. 5.1 below:

While there may be details within this schematic which we might question, or add to, it does provide a useful separation of the underlying causes of deforestation, and what Geist et al. term "proximate causes" but which we would suggest could equally be interpreted as some of the *symptoms* which are manifested when a deforestation

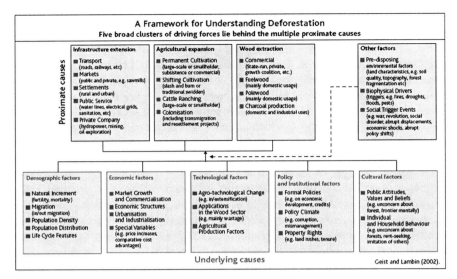

Fig. 5.1 Deforestation: Causes and Symptoms from Geist and Lambin (2002)[6]

dynamic is in progress. This might be seen as a distinction without a difference, but we suggest this is definitely not so: The primary use of identifying symptoms, in any living system, is to assist in the diagnosis of the disease or disorder present. It may not be possible to treat the disease in some cases, in which case relief of the symptoms is the only course of action. However, if the disease can be treated effectively, then priority must be assigned to doing so. So, as far as possible, we need to know what we are dealing with.

In the 1990s, as concern about deforestation, especially tropical rainforest loss, began to rise, some analysts sought quantitative, correlative approaches to linking observed deforestation to causal factors. Kaimowitz and Angelsen (1998) reviewed a large number of these quantitative models investigating the causes of deforestation. The authors conclude that rising agricultural prices, lower wages, reduced off-farm employment and more road construction seem to lead to higher deforestation. Less certain are the impacts of technological change, tenure security, agricultural input prices, and macro factors: poverty reduction, population growth, economic growth and debt. However, the authors caution that the models results, based on national or multi-national level data, can suffer from data and methodological problems. They suggest that more localized and field-based studies could yield more reliable results.

This is sound advice: There is a great deal of literature available on the subject of *drivers of deforestation* and some of it does tend to equate the loss of forest in a given situation – often at very large, even national, scale – with the pace of growth

---

[6] American Institute of Biological Sciences. Used by permission. All rights reserved.

of an activity, or group of activities, *believed* to be the main cause of deforestation. The problem with this approach is that it assumes, implicitly, that a reduction in the force of a given driver will lead to a corresponding drop in the level of deforestation present: in other words that there is a direct cause-and-effect mechanism operating.

However, a correlation does not imply causality: it is apparent from many situations in the field that removal of a given driver of deforestation simply creates a vacuum into which another deforesting activity will soon move. Secondly, what appears to be a driver, simply because that particular activity is present on a given deforesting site at the time of observation, does not necessarily mean that that activity was the main cause of forest loss. It may well be that a number of activities which have continuously degraded the forest area in question may have been in operation there for years, or decades, and that the activity which eventually has replaced the forest (or is in the process of doing so to whatever remains) may not have been the main agent of forest loss – it may simply have been the last one to arrive in that location.

Even if we know more than the simple correlation – for example, if we know that government plans have identified forested areas for conversion to commercial scale crop plantation, years ahead of the actual final clearing and felling – this *still* does not prove that the deforestation is therefore attributable to the crop project. It may simply have been the result of a government agency bowing to the inevitability of losing a forest under heavy logging and other activities already present, by suggesting that since the forest area is already degraded, it would best be converted to some other profitable purpose. Or it might signify that the government forest agency itself wants this result, or is under political pressure from other national ministries and also from local governments and groups, to let this happen.

In other words, if we do not understand the motives and incentives which are driving the behaviour of all groups – official, local, legitimate and illegitimate – then we cannot know what combination of incentive, disincentive, regulatory and capacity building measures will work to reduce deforestation effectively in that particular site.

Until this in-depth work is done (if, in fact, it ever is) there is one line of inquiry we can follow which might at least give us some idea of what to do about deforestation, and that is to address the question of forest value. As we will quickly discover, in the following chapters, that is by no means a straightforward matter either, but it will provide some insights into what might work, and what will not.

References

ABIEC (Association of Brazilian beef exporters) undated. www.abiec.com.br Accessed 2008
Angelsen A, Kaimowitz D (2001) Agricultural technologies and tropical deforestation. CABI, Wallingford, UK
Chomitz KM (2006) At Loggerheads? Agricultural Expansion, Poverty Reduction and Environment in the Tropical Forests. Policy research review draft report, World Bank
DFID (2004) Money Laundering and Illegal Logging. Department for International Development, Government of the United Kingdom. http://www.illegal logging Access date uncertain

DFID (2007) Crime and persuasion: tracking illegal logging, improving forest governance. Department for International Development, Government of the United Kingdom
Dietz T, Dolsak N, Ostrom E, Stern P (2002) The drama of the commons. National Academy, Washington, DC
Friends of the Earth – Amazonia Brasileira (2008) Friends of the Earth – Amazonia Brasileira. The Cattle Realm – A new phase in the livestock colonization of Brazilian Amazonia (2008) http://www.amazonia.org.br/english/guia/detalhes.cfm?id=259673&tipo=6&cat_id=83&subcat_id=401. Accessed 2009
Geist H, Lambin (2002) Proximate causes and underlying driving forces of tropical deforestation. BioScience 52(2):143–150. © 2002, American Institute of Biological Sciences. Used by permission. All rights reserved
Global Timber undated http://www.global timber.org
Grunwald M (2008) The clean energy scam. Time Magazine, March 27
Hardin G (1968) The tragedy of the commons. Science 162(3859):1243–1248
Kaimowitz, D. & Angelsen, A (1998). Economic Models of Tropical Deforestation: A Review, Center for International Forest Research, Bogor, Indonesia
Kartodihardjo H, Supriono A (2000) The impact of sectoral development on natural forest conversion and degradation: the case of timber and tree crop plantations in Indonesia. CIFOR Occasional Paper No 26(E), January
Koh & Wilcove. (2008). Is oil palm agriculture really destroying tropical biodiversity? Conservation Letters (2008). Blackwell Publishing, Inc.
Lawrence F (2006) Should we worry about soya in our food? The Guardian 25 July 2006. http://www.guardian.co.uk/news/2006/jul/25/food.foodanddrink. Accessed 2008
Maturana J (2005) Economic costs and benefits of allocating forest land for industrial tree plantation development in Indonesia. Center for International Forest Researcwh, CIFOR, Bogor, Indonesia
MongaBay (2009) Tropical rainforest: saving what remains http://rainforests.mongabay.com/1009.htm. Accessed 22 Apr 2009
Nepstad D, Stickler C, Almeida O (2006) Globalization of the Amazon soy and beef industries: opportunities for conservation. Conserv Biol 20:1595–1603
New Energy Finance (NEF) (2008) Brazil Ethanol Quarterly Outlook. 10 April 2008
OECD (2008) Environmental Outlook. Summary. http://www.oecd.org/dataoecd/29/33/40200582.pdf. Accessed 2009
Ostrom E (1990) Governing the commons: the evolution of institutions for collective action. Cambridge University Press, New York
Stern NH (2006) Stern review: the economics of climate change. HM Treasury, Government of the United Kingdom, UK
Thow A, Warhurst A (2007) Biofuels and Sustainable Development. Maplecroft. 2007
United States Department of Agriculture. (2006) Palm Oil Continues to Dominate Global Consumption in 2006/2007. http://www.fas.usda.gov/oilseeds/circular/2006/06-06/Junecov.pdf
United States Department of Agriculture (April 2008) Livestock and Poultry: World Markets and Trade. http://www.fas.usda.gov/dlp/circular/2008/livestock_poultry_04-2008.pdf
USMEF (2007) Beef Export Forecast. http://www.cattlenetwork.com. Accessed 2009
Vera-Diaz M, Schwartzman S (2005) Carbon offsets and land use in the Brazilian Amazon. In: Moutinho P, Schwartzman S (eds) Tropical deforestation and climate change. Amazon Institute for Environmental Research, USA Environmental Defense, Belém, Pará, Washington, DC
Volpi G (2007) Climate Mitigation, Deforestation and Human Development in Brazil. Human Development Report. United Nations Development Program, 2007/2008
VROM Think Tank International Affairs Secretariat (2008) Stimulating sustainable trade aiming at a joint government – business approach addressing non-trade concerns, 22 Feb

Part III
Sustainability and Valuation of the Forests

Chapter 6
Sustainability Versus Ideology in the Forests

Abstract This chapter begins with an historical perspective on the shifting paradigm of environmental stability. It reviews the original conception of natural resource depletion as argued by Malthus, and his inheritors in the twentieth century: the Club of Rome, Ehrlich and others. The case for comprehensive environmental collapse was not well argued at this point, and as a consequence was set upon by economists and growth protagonists. Nevertheless, it may be that as environmental and natural resource constraints become more evident at the global scale – as seems to be the case for the climate change scenario – some of this earlier concern may come to be seen as premature, rather than fundamentally wrong.

While there exists a great deal of work on the microeconomics of forest use and value, there has been relatively little on linking forest outcomes (or indeed environmental ones) to macroeconomic policy: in Chapter 7 of this book, further exploration of this issue is undertaken. Concerns of this nature have not featured strongly in the raft of multilateral agreements on global environmental stability and forests: these agreements and their implications are reviewed briefly in the chapter.

One of the reasons why broad linkages of cause-and-effect in forests has not always been addressed is due to the presence in many of the international forests constituency groups involved of highly ideological and restrictive interpretations of what needs to be done. The chapter closes with an illustrative case study of this, based on the World Bank's forests policy.

The issue posed in the title to this chapter above can only be dealt with after the much larger subject of the sustainability of the full set of natural resources upon which humanity depends for its long term survival and progress is addressed. At this level of aggregation, forests are treated in discussion immediately following as an element (albeit an important one) in the rapidly evolving global sustainability picture. Once this broader context is established, the specific issues defining the forests situation will be considered, later in this chapter.

6.1 Global Environmental Sustainability: The Shifting Paradigm

One indicator of the speed and extent to which the global sustainability paradigm has shifted in relatively recent times has already been demonstrated in this book in Chapter 2, using the example of Lawrence Summers' heroic optimism as to the sustainability of the carrying capacity of the Earth in the face of human activity, as he saw things in the early 1990s, and contrasting this with the world view demonstrated now by Nicholas Stern (Stern, 2006) who also occupied the position of Chief Economist with the World Bank, as had Summers, before moving on to becoming a leading figure in the economics of climate change.

Stern's intellectual trajectory on the issue of sustainability mirrors a growing appreciation in his profession of the fragility of many environmental systems which harbour the natural resources upon which many of the assumptions behind conventional notions on economic growth depend. There is a case for arguing that many of the most serious ramifications of mistakes that have been made in the design and implementation of the broad economic policies and growth strategies that economists have produced have been manifested in the natural resources sectors. Time and again, implicit assumptions about the continued availability of arable land, the resilience of that land under intensive farming, the supply of logs and other forest outputs, the sustainability of fisheries and, most pressingly in recent years, the viability of water supplies, have eventually been proven to have been too optimistic.

Malthus, the Club of Rome, and the environmental Kuznets curve.

To understand why environmental sustainability has taken some time to rise to significance in economists' concerns, it is necessary to briefly examine the history of the sustainability subject. Such a review indicates that, if nothing else, the main protagonists of the idea that global sustainability is at risk have not always proven to be the best advocates of their case, and at times almost seem to have concluded that the best way to market an unpleasant truth is in fact to embellish it with overblown rhetoric and horror stories sufficient to render it completely unpalatable.

This in turn produced a backlash of scepticism from economists, based largely on their conviction – perhaps even ideology – that the self correcting nature of markets would guarantee that the limits to economic growth will always be less significant and non-linear than is commonly imagined or argued by the growth pessimists.

6.1.1 Malthus: The Original Prophet of Economic Doom

The Reverend Thomas Robert Malthus (1766–1834) wrote broadly on a number of economic and social issues, including economic rent and the theory of money, and engaged in vigorous debate with the eminent economist, David Ricardo, on the theory of value. He is best known for his *Essay on the Principle of Population*[1], first published in 1798, and re-issued in much expanded form in 1803, in which he argued that:

The power of population is so superior to the power of the earth to produce subsistence to man, that premature death must in some shape or other visit the human race ...

In very reduced form, Malthus' reasoning was that, if not constrained in some way, population will increase at a geometric rate, whereas food supply can only grow at an arithmetic rate. The gloomy consequence of this was that sooner or later, population growth would outstrip the availability of sufficient food supplies.

Malthus has been criticized (from hindsight) by many in the ensuing two centuries (Engels described his ideas as "the crudest, most barbarous theory that ever existed..."). It is interesting to note, however, that since his essay was published, global population growth did grow at something like an exponential rate for a considerable part of the period (although not in Britain and Europe, which were the focus of his concerns). What Malthus did not conceive of (and could not have, given the information available to him at the time) was the massive expansion of agricultural production that has been attained over this period. Malthus did influence important figures such as Darwin and Karl Marx, and is credited by some as having motivated William Pitt the Younger's attempted amendment of the Poor Laws.

6.1.2 The Inheritors of Malthus

The best-known example in the modern era of neo-Malthusian analysis is *The Limits to Growth* (Meadows et al 1972), sponsored by the Club of Rome. The analysis of world economic developments in this study contained one of the earliest examples of application of large computer capacity to a multiparametric projection model. Using an approach called system dynamics, it linked growth patterns and demographic trends to estimates of natural resource constraints, via a series of feedback loops, which allowed given actions to be linked to their impacts on surrounding environmental conditions, and thence to the influence of that on subsequent actions, and so on.

The approach demonstrated fealty to the basic Malthusian approach, by combining exponential projections of factors such as population growth and consumption, with fixed limits on basic parameters such as land, and other non-renewable resources. Not surprisingly, given this, it concluded that within 100 years, with no major change in physical, economic and social relationships, society will run out of critical non-renewable resources, leading to massive consumption overshoots, and then economic collapse. The study concluded that piecemeal approaches to solving the

---

[1]Published under World's Classics Series, 1983, Oxford University Press.

individual problems arising in the projected scenario will not succeed. For example, a doubling of the assumed level of resources does not change the outcome – economic collapse – but alters the cause of this to the polluting impacts of high rates of industrial growth, leading to massive pollution and the associated impacts on populations. Even if the pollution problem were to be solved, population growth would continue to accelerate to the point where food supplies are overwhelmed. Overshoot and collapse can, according to the study, only be avoided by limits on population growth, control of pollution, and (significantly, in view of subsequent reactions to the theory) serious reductions in economic growth.

The *Limits to Growth* was part of an explosion of doomsday predictions in the 1970s; another piece of work which gained great prominence at about the same time was Paul Ehrlich's *The Population Bomb* which predicted that in the 1970s and 1980s, hundreds of millions of people would starve to death, basically because of consumption outstripping supplies at a global scale. As a somewhat ironic aside on this, in 1980 Ehrlich and two other academics undertook a wager with Julian Simon (1932–1998), a well-known business economist who had expressed doubts about the dire nature of the resource depletion projections in the work of Ehrlich and others. Simon bet that the real (i.e. inflation-adjusted) price of five commodity metals would decline in real terms (suggesting relative abundance, not scarcity) by 1990. This proved to be the case, and Ehrlich paid up in 1990. Had the bet been based on the projection year 2008, however, Ehrlich would have won, because by that year, the real prices of all but one of the metals chosen had risen[2], based on the 1980 base year (Bio-Law website 2008).

6.1.3 Growth Protagonists Push Back

In 1975, Herman Kahn and others (Kahn et al. 1976) provided an early repudiation of the findings of the Club of Rome study, developing on the argument that technological developments would push back limits on natural resources. They suggested that population growth was in fact following an S-shaped curve, rather than a consistently exponential one, and that rates of population change would eventually decline to zero. There is little doubt that this is the case, over the relatively long term: present population projections indicate a population level for the earth of around 9.5 billion by 2050. This represents a significant slowing of the rate of change of population by that year, although it is still the case that absolute additions to the population total between now and then will be higher than the numbers added in previous equivalent periods.

---

[2]Since then the prices have collapsed but increasing consumption may bring them back to the growth path, however, year 2008 may prove to be a short-term peak in the evolution of metal prices.

As suggested earlier, there have been numerous negative responses to the *Limits to Growth* and related studies ever since its release, from the analytical, through the essentially ideological (exemplified in the quotation from Lawrence Summers cited earlier), to the extremist. For an example of the latter, see an article[3] which among other things, suggests that a burgeoning population should not be of concern because

> if every one of the 6 billion of us resided in Texas, there would be room enough for every family of four to have a house and 1/8 acre of land – and the rest of the globe would be vacant.

and goes on to assure us that

> the "energy crisis" is now such a distant memory that these days oil is seen as the cheapest, not the most expensive liquid on Earth.

6.1.4 The Environmental Kuznets Curve

In the early 1990s, articles began to appear which adapted an economic theory developed by Simon Kuznets (1955) to the question of environmental degradation. Kuznets' original work was an analysis of the relationships between income growth and inequality, and it argued that if a measure of income equality (the ratio of income going to the top earning 20% of the population compared to the bottom earning 20%) was plotted on the vertical axis against income per capita on the horizontal axis, then an inverted U-shaped curve would result (Fig. 6.1).

The explanation offered for this pattern was that in the early stage of economic development, there is a premium on investment capital for the development of physical resources, so those capable of saving and investing most are rewarded.

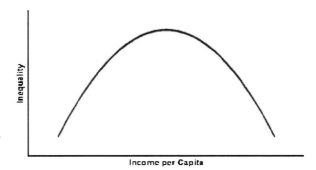

Fig. 6.1 The Kuznets relationship between income and inequality

---

[3]Stephen Moore *Defusing the population bomb* Washington Times (October 13, 1999). Excerpted from an article that appeared in *The Washington Times*, October, 1999. (c) The Cato Institute. Used by permission.

Later in development, as workers shift from rural areas and agricultural activity to the higher paying urban jobs being generated, the proportion of total income going to their wages rises, in relative terms.

The World Bank (1992) published some analytical work which plotted ambient air and water quality against GDP per capita figures and discovered a similar relationship as that indicated in the original Kuznets curve – in other words, an initially strongly rising level of pollution in air and water as economic development and per capita incomes begin to grow, followed by a lessening and eventual reversal of rates of pollution growth as economic progress raises public awareness and allows such problems to be addressed. Later applications of this approach to automotive lead emissions, deforestation, greenhouse gas emissions and toxic waste have revealed similar patterns, although not always as clearly defined as in the original case studies.

Grossman and Kreuger (1994) used data assembled by the Global Environmental Monitoring System to examine the relationship between various environmental indicators (concentrations of urban air pollution; measures of the state of the oxygen regime in river basins; concentrations of fecal contaminants in river basins; and concentrations of heavy metals in river basins) and the level of a country's per capita income. The authors found no evidence that environmental quality continues to deteriorate steadily with economic growth. Rather, for most indicators, economic growth brings an initial phase of deterioration followed by a subsequent phase of improvement – i.e. the Kuznets result. They also found that the turning points on the curves for the different pollutants vary, but in most cases came before a country reached a per capita income of $8,000.

6.1.5 Taking Stock

Notwithstanding whatever satisfaction Professor Simon may have derived from his wager with Ehrlich et al., the methodological controversies generated by the *Limits to Growth* and related studies in the 1970s continue in various forms in the literature until the present. Defences of the approach have been produced since the original publication, by the Club of Rome and others (see Suter 1999; van Dieren 1995), and in 2004 some of the original authors issued a re-appraisal (Meadows et al. 2004). Nevertheless, it can be argued that, for most of the period since these studies were released, the adverse reactions of economists and others (including many involved in development assistance programs in poorer countries, who did not find the strictures on growth prospects called for in the studies helpful) have probably had more influence on economic policy and outcomes than the recommendations of the studies themselves.

It remains an open question at this point as to whether this will remain the case, given the advent of peak-oil pessimism, climate change issues and related concerns that suggest that limits to growth are growing more visible on the near horizon. That august journal of record of market economics, *The Economist*, recognizes the need to act, but clearly does not believe that Malthus provides the guide as to how[4]:

> A new form of Malthusian limit has more recently emerged through the need to constrain greenhouse-gas emissions in order to tackle global warming. But this too can be overcome by shifting to a low-carbon economy. As with agriculture, the main difficulty in making the necessary adjustment comes from poor policies, such as governments' reluctance to impose a carbon tax. There may be curbs on traditional forms of growth, but there is no limit to human ingenuity. That is why Malthus remains as wrong today as he was two centuries ago

Paul Krugman[5], the Nobel laureate, liberal economist and columnist, and an original critic of *Limits to Growth*, argues that the methodology of the study – especially some of the rigid assumptions built into the original projection model provided to the *Limits* study by Professor Jay Forrester – was flawed: Forrester did not, according to Krugman, have a grasp of the empirical evidence on economic growth, and his projections suffered accordingly.

However, even though Krugman is unconvinced by *Limits to Growth*, this does not mean he is optimistic about the outlook for global resources and environmental sustainability: he notes that progress on energy technologies, and indeed humanity's ability to manipulate the physical world in general has been disappointing, with the implication that reliance on these in the future to solve our problems of growth and sustainability would be unwise.

Thus, the irony is that the general messages of depletion and decline that have been sounded from the ideas of Thomas Malthus in the early nineteenth century through to the Club of Rome and Paul Ehrlich (1968) in the 1970s, and widely scorned by economists and others since (for good reasons) are now seeming considerably more imminent, albeit for reasons, and in ways, that Malthus and his more recent adherents could never have imagined.

Gus Speth (2008) shares Krugman's reservations about modern capitalist nations' ability to deal effectively with resource depletion and environmental degradation and their consequences:

> There are many good reasons for concern that future economic growth could easily continue its destructive ways. First, economic activity and its enormous forward momentum can be accurately characterized as "out of control" environmentally, and this is true in even the advanced industrial economies that have modern environmental programs in place. Basically the economic system does not work when it comes to protecting environmental resources, and the political system does not work when it comes to correcting the economic system

The above review will go some way towards explaining why there is considerable attention paid in this chapter to the views and approaches of economists, in comparison to other perspectives that are relevant.

---

[4]*Economics focus*, May 17 © The Economist Newspaper Limited, London 2008. Used by permission.
[5]Krugman, Paul, *The Conscience of a Liberal*, New York Times, May 6, 2008.

6.2 Forests and the Broader Economy

6.2.1 Applying the Environmental Kuznets Curve to Forests

If the Kuznets result as described by Grossman and Kreuger for various environmental indicators were to apply to forests, for a middle-sized forestry developing country, then the outcome would look something like Fig. 6.2 below:

David Stern et al. (1996) critically examined the application of the environmental Kuznets curve in a number of studies, and showed that when some of the assumptions on feedback of environmental quality into production, and on trade impact neutrality are not met in practice, problems with estimating the parameters of the curve arise. Some of the estimates made in the literature that further development will reduce environmental degradation are dependent on the assumption that world per capita income is normally distributed, when in fact median income is far below mean income.

The authors found when aggregating the results of various studies through time, global forest loss stabilizes before 2025, but tropical deforestation – which has been the source of primary concern in relation to deforestation – continues at a constant rate throughout the period. This serves to emphasize the fact that the time frames over which the Kuznets effect operates are important.

Contreras-Hermosilla (2000) cites multi-country studies by Cropper and Griffiths (1994) and Panayotou (1995), and a study of Malaysia by Vincent et al. (1997), which all indicate the Kuznets inverse U relationship between per capita income and the rate of deforestation holding. However, he also cites another case study by Sunderlin and Pokam (1998) which indicates that in some cases where a country's per capita growth rate actually goes into reverse, this does not necessarily mean that the rate of deforestation will also reverse according to the Kuznets pattern. Contreras-Hermosilla cautions that:

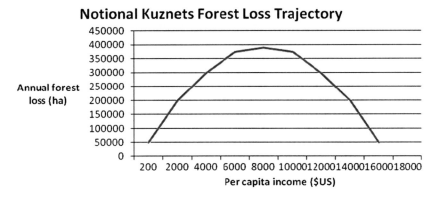

Fig. 6.2 Notional Kuznets Relationship between Forest Loss and Income

> Explanations attributing an overwhelming driving force to one variable such as income growth are too simplistic .... The numbers and complexity of underlying causes of forest decline calls for caution. It is not possible to find unambiguous cause-effect linkages that would have a universal application. Rather, specific situations must be studied in detail and remedies must also be specific

A cross-sectional study by Culas and Dutta (2002) which examined the relationship between deforestation and factors such as per capita income, economic growth and agricultural growth showed the inverted-U pattern for Latin American and African countries, but a U-shaped relationship for Asian countries.

This suggests that forests are unlikely to disappear completely – so long as rising incomes globally can be maintained – but, obviously, poorer countries will lose more forest from this point before any Kuznets effect will be felt. While the Kuznets curve is useful conceptually, it is clear that it often operates over long time frames: where concerns for forest sustainability are serious and immediate, the reason is rapid deforestation, and if the country concerned still has relatively low median income, then waiting for the Kuznets income effect to reduce that deforestation would certainly not be an effective strategy. This is especially the case given that most of the empirical studies of the environmental Kuznets curve cited above – whether for deforestation, or more generally for environmental improvements – show that the point of inflexion in the Kuznets curve does not occur until relatively high per capita incomes are reached. Contreras-Hermosilla's injunction to apply case-specific analyses when studying causes and remedies for deforestation therefore seems apposite.

The environmental Kuznets curve serves a useful purpose conceptually, demonstrating that in aggregate terms, and over long time frames, policies which impact upon per capita income growth at national scale can be expected to exercise differential effects on forests (eventually) as growth proceeds, but of itself has limited application to positive policy development to address forest sustainability in particular cases. It does serve, however, to point up the need for a stronger linkage between large economic variables and policies, and decision-making on what is to be done in forests.

6.2.2 *Forests in the Broader Economy*

A general critique which underlies the case for a re-appraisal of forest intervention by the international forests constituency is that much of what it has done in forests in the past has been determined by an insider perspective – essentially a process of sectoral specialists identifying what they see as problems specific to forests and those involved in their use, with little or no reference to the wider economic framework in which the sector is located and which may be the predominant source of much of the dynamic which is driving change in the forests – for better or worse. It would not be fair, we suggest, to attach blame for this to those who have been most involved. The reality, until very recently, has been that it has simply not been possible to engage a broader economic and social interest in this subject, in many of the developing countries where large natural forests remain, especially in cases where there are strong vested interests in the highest political levels, which make corrective action difficult, particularly if there are no democratic checks and balances in effect.

To date, there has been plenty of useful microeconomic analysis of forest management and use for the many goods and services forests can provide, and considerable work done on the domestic and international trade aspects of forests products. There has been relatively little examination of the impacts of broad economic policy – including macroeconomic policy – on forest outcomes. As we will see in Chapter 7, there has been considerable debate in forests literature as to whether or not these impacts are likely to be significant: the broad conclusion we have drawn from this is that they certainly can be, but the manner and even the direction in which this manifests from one location to another is highly variable. There has been little specific field work done on how the broad instruments of macroeconomic and inter-sectoral growth policy could be developed or refined so as to address issues of perverse incentives and unsustainable outcomes in forest locations, even in cases (in fact, the majority of nations) where governments proclaim adherence to general principles on sustainable management and equitable development of forest resources – often to the extent where these are enshrined in national constitutions and basic law.

This situation can be seen as emblematic of the intellectual divide between scientists, environmentalists and ecologists and others who study aspects of the physical world on the one hand, and economists (sometimes joined by sociologists and others) who are focused on human progress. Scientists are used to natural systems which can change and evolve in different directions. Tides move in and out, new species are created, while others go extinct – there is no primary direction in all of this. In the case of economists there is a primary direction behind all matters they consider, and it is the direction of economic growth. Unless things are moving in that direction, the economist will see developing instability, leading eventually to social chaos and political breakdown. Where a scientist will see the logical response to a human activity which is causing some adverse natural resource development as being simply to halt or reform that activity, an economist – if indeed he or she perceives the seriousness of the problem at all – will see the primary necessity as being the need to transform the growth dynamic implicit in that activity to a more sustainable one, even if that is less immediate and effective in arresting the adverse trend itself.

A consequence of this disjunct between economic and scientific thinking has been that it has been relatively difficult to interest economists in the immediacy of the linkage between economic growth and environmental degradation, and therefore of the need to utilize the full armoury of macroeconomic policy reform to address environmental degradation. In the following chapter in this book, we will show how the climate change issue has the potential to change all this.

By the same token, scientists have often failed to see the relevance of reforms and adjustments in rather arcane economic variables and incentives – often far removed from the immediate site of the environmental problems of most concern – that can be of very real consequence. In perhaps over-simplified terms, economists frequently have relatively little understanding of the real nature of the linkage between economic development and environmental degradation, while scientists have equally little understanding of how remedies from the economists' toolkit might effectively address the observed problem.

We will return to this subject in Chapter 7 of this book, where we will review the options for financing sustainability in forests that potentially will be opened up by climate change.

6.2.3 Financing Sustainability in Forests Has Been Inadequate

Had the developed countries been as concerned as their utterances on the subject of global deforestation have suggested, then the disjunct noted above might not have mattered: to put it somewhat crudely, flows of finance could have bridged the gap, by adding value to the conservation/sustainability approach. However, while flows of overseas development assistance (ODA) into forests related activities have increased markedly (see Table 6.1 above) it has virtually never risen above $2 billion per annum in total. As will be discussed further in the following chapter, given the size and lucrativeness of the forest resource, and the lands they occupy in many rainforest nations, this level of funding has not been adequate to provide a significant disincentive to deforestation.

To place the amounts in this table in some context; total global ODA for all purposes in 2004 (in 2004 $US) was $80 billion. By 2007 this had risen to $91 billion, but this was far short of the amount donor nations had committed to by that year at the Millenium Summit ($115 billion), and it seems likely that the target of $130 billion for 2020 will also not be met, by a wide margin. On this basis, ODA to forests has not yet risen above 2% of all ODA funding. The proportion of total lending by multilateral banks going to forests purposes has fluctuated a good deal, but around the same figure of about 2% of the total.

Table 6.1 External financial flows to forests (Simula 2008)

Source	2000–2002	2005–2007	Change
	USD mill. at 2006 exchange rates and prices		%
ODA[a]			
– Bilateral	929.1	1,078.7	+16.1
– Multilateral	335.0	806.7	+140.8
Total	1,264.1	1,885.3	+49.1
Private sector			
– Foreign direct investment	400.0	516.0[c]	+29.0
– Other private financing	–	–	Increase
NGO, philanthropic and others	–	–	

[a]Appendix 4.X
[b]UNCTAD 2007
[c]2003–2005

More recently the developed countries have put forward a raft of new funds and mechanisms which are either focussed on, or which include, substantial programmes for forestry:

- The World Bank's Forest Carbon Partnership Facility, supported by several international donor groups, with a proposed budget of $300 million of which $170 million has been pledged.
- The World Bank's Strategic Climate Fund, which intends to finance adaptation to climate change but has a specific interest also in reducing deforestation, has funding commitments from the G8 group of nations of $ 6 billion and will include a Forest Investment Programme.
- The Norway Forest Fund, which has committed $2.8 billion over 5 years from 2008.
- The Congo Basin Fund, supported by Norway and the United Kingdom, with funding of $195 million.
- The Japanese Government's Cool Earth Partnership designed to support adaptation to climate change and access to clean energy, with some forest interest, to run for 5 years, allocating $2 billion per year from a $10 billion fund.
- The Australian Deforestation Fund, aimed at reducing deforestation in the Southeast Asia region, with funds of $A200 million.
- The German commitment of 500 € million a year for biodiversity.
- The suggestion by the European Commission for the creation of a Global Forest Carbon Mechanism (GFCM) (although some suggest this may be compensation for avoided deforestation most probably being excluded from the European Emissions Trading Scheme until at least 2020).
- Brazil's Fund for the protection of the Amazon rainforest has received a commitment for an initial $130 million from Norway (drawn from the Norwegian Forest Fund) and Guyana has offered to place its forest under international stewardship in return for compensation for development opportunities foregone.

6.3 Multilateral Agreements on Global Environmental Sustainability

While the developed countries have been less than generous in funding efforts to promote sustainability and conservation in rainforest countries, they have been more forthcoming in their willingness to participate in forums, discussion, treaties and the like which touch upon the environmental sustainability issue.

The economic debates on the limits to growth – essentially the question of whether economic growth is the main problem, or the main solution, for natural resource sustainability, as discussed earlier in this chapter – continued through the 1970s, 1980s, and 1990s, and have if anything intensified as the realities of the climate change issue have begun to dawn, the issue of global environmental sustainability has loomed increasingly large on the international stage. In the public mind in developed countries, environmental sustainability has increasingly become the watchword by which real progress in human development and economic growth are evaluated. There have been innumerable conferences, colloquiums and workshops on this subject; large numbers of research and academic programs have been founded upon it, as public interest in the general subject has intensified.

The changing nature of international and intergovernmental discussion and negotiation on this subject can be traced through review of the five major global gatherings on environment of recent decades.

6.3.1 The Stockholm Agreement

In 1972, 113 nations convened at the United Nations Conference on the Human Environment, in Stockholm. This was the first genuinely global gathering on the environment. It considered:

- Human impacts on the environment
- Social and economic development and population growth
- Related issues for developing countries, and for international assistance in these areas
- The role of government in developing and managing environmentally friendly development
- The potential contributions of technical development and education to addressing environmental issues

Given the early date of this conference, the coverage of issues was quite broad, and the linkage between environmental degradation, persistence of poverty, and government policy and practices was recognized. While few specific demands were made on participating nations to implement monitorable, quantitative changes in relation to environmental management, this conference did at least serve to establish a broad mood of concern about global environmental issues. A number of international, national and non-governmental organizations focused on the environment were formed as a result of it – notably, a new United Nations agency for the environment; the United Nations Environment Programme.

6.3.2 The Brundtland Report

The World Commission on Environment and Development – usually referred to as the Brundtland Report, after its chair, Gro Harlem Brundtland – was established by the United Nations in 1983, and submitted its report to the general Assembly of the UN in 1987.

The Brundtland report reflected the views of a majority of its contributors, and concluded that the most critical global environmental problems were primarily the result of the enormous poverty of the developing South, and unsustainable consumption of natural resources by the rich Northern nations. The report is generally credited with bringing the term "sustainable development" into common usage, in its call for a global strategy that would link development and environmental protection. It provided a simple and now widely used definition of the concept:

> Sustainable development is development that meets the needs of the present, without compromising the ability of future generations to meet their own needs.

This inter-generational consideration in development has become an important element in subsequent deliberations on environmental sustainability. Almost every definition of sustainable forest management reproduces this inter-generational stricture, regardless of how differently the definitions play out beyond that point. At the discussion of the report at the general Assembly, it was resolved that the matters raised by the Brundtland Report should become the subject of a large and comprehensive international and intergovernmental conference and development.

6.3.3 The Rio Earth Summit

More formally known as the UN Conference on Environment and Development, this international meeting took place in June 1992, In Rio de Janeiro. It was (and remains) by far the largest and most ambitious international gathering on the global environment ever undertaken. Over 30,000 delegates attended, representing various interests from 172 countries; more than 2,000 non-governmental organizations took part; and about 100 heads of state participated directly in the meetings. The broad theme of the conference was that a significant change in the attitudes and activities of people in their relationship with the natural environment was now necessary. The linkage between poverty and environmental degradation made in the Brundtland report was reprised and developed, as were the damaging environmental consequences of affluence and excessive consumption.

The Rio Earth Summit produced international agreements, understandings or declarations on:

- Protecting the biodiversity of the planet
- Addressing climate change
- Managing the world's forests
- The status of the environment and economic development
- An action agenda for governments to implement to address a broad group of environmental and sustainability concerns (known as Agenda 21)

Copious documentation of the proceedings and agreements was produced, focusing on the requirements of Agenda 21 (a broad ranging set of global objectives, covering the gamut from economic and social improvements, through conservation and management of resources, to effective participation by major groups in civil society), the Rio Declaration on Environment and Development, The Statement on Forest Principles, the UN Framework Convention on Climate Change, and the UN Convention on Biological Diversity. Follow-up mechanisms established within the United Nations organization included the Commission on Sustainable development, the Inter-agency Committee on Sustainable Development, and the High Level Advisory Board on Sustainable Development.

Whatever else may be said of the Rio process in retrospect, it cannot be denied that in its wake, the broad issue of environmental sustainability was well and truly established at the top of the international conference agenda for quite some time.

6.3.4 The Kyoto Protocol

The Kyoto Protocol to the United Nations Framework on Climate Change is, in effect, an international agreement on measures to address climate change which amends and clarifies the provisions and agreements on this issue produced at the Rio summit. The objective of the Protocol is the stabilization of carbon dioxide and other greenhouse gases in the atmosphere at a level that will prevent continued dangerous human interference with the global climate system.

Negotiations of the Protocol began at the first international conference on the subject, held in Kyoto in 1997. Initial negotiating conditions were that at least 55 countries must sign the convention, and that in total the countries then involved must account for at least 55% of global carbon dioxide emissions, as estimated in 1990. Iceland's ratification of the convention in 2002 brought the number of participating countries involved to 55, and Russia's ratification in 2004 brought the total emissions represented by the signatories to more than 55%. The treaty came into force formally in February, 2005.

Events since then, on the forests side of things, have moved uncharacteristically quickly, with the UNFCCC COP 13 meeting in Bali, where important initial commitments to reducing deforestation and degradation were made, and the ambitious agenda which has been set for initiation of the REDD process at the COP 15 meeting in Copenhagen in late 2009. We will return to these subjects in Chapter 7.

6.3.5 The World Summit on Sustainable Development

Convened in 2002, 10 years after the Rio Summit (and often referred to as Rio +10) this gathering was organized to review progress towards sustainability since Rio, with a focus on implementation of existing agreements and undertakings. In the official language of the organizers, the intention of the 2002 summit was to review the successes and failures of countries in meeting their commitments made at Rio in a frank manner, to reinvigorate the global commitment to sustainable development, and to deepen the global commitment to sustainable development through a new "global compact", and bring a new spirit into the environmental debate.

There was a general consensus among UN member states that the Agenda 21 principles agreed on at Rio in 1992 should not be renegotiated. It was also agreed amongst most participants that the primary focus of the Summit should be on "poverty, development and the environment". Thus, poverty and underdevelopment remained on the agenda as the fundamental threats to environmental security and sustainable development.

In general terms, the conference was intended to be lower profile: it would be a little lighter on production of sweeping international treaties and agreements than was usual, instead emphasizing arrangements for better cooperation on existing goals (under Agenda 21 and other agreements reached at Rio), identification of new challenges and opportunities that had emerged since Rio, and further consideration of the issue of balance between social and economic development and environmental protection. Some refinement or quantification of details for these goals was to be undertaken, so that more definitive monitoring of progress could be done.

In the event, the Plan of Implementation from this summit ran to several chapters, each containing dozens of recommendations. In some cases, these revisit old ground, such as reiteration of existing commitments made under the World Trade Organization ministerial meetings, and "re-committing" developed nations to dedicating at least 0.7% of their national income to development assistance – a goal that was originally set in 1970, and which has been attained by very few countries since then. Specific new targets were agreed upon – for example, to reduce the numbers of people without access to clean drinking water by 50% by year 2015, and to cut the rate biodiversity loss "significantly" by 2010. More than 300 voluntary partnerships for various aspects of sustainable development and conservation were established at the conference.

6.4 A Brief Look at Multilateral Involvement in Forests

The broad multilateral dialogue on environmental sustainability has generated a large number of consequential initiatives and actions in global forests, particularly from the time of the Brundtland Report onwards: implementation of the Tropical Forestry Action Plan; formation of the International Tropical Timber Organization; The World Commission on Forests for Sustainable Development; a series of intergovernmental dialogues on forests and sustainability running through to the present incarnation, the United Nations Forum on Forests; the outflow on forests from the Rio Earth Summit; and many more. Much has been written and said about these, and we can do no more here than briefly overview some of the seminal developments, and in some cases draw our own conclusions as to their implications.

Not surprisingly, these large international activities on forests have been the focal point for much of the debate and conflict which has gone on in the international forests constituency. Policy responses and programme adaptations in response, within the various multilateral and bilateral agencies involved in forests, and in NGOs and other entities involved, have been vigorous and in some cases traumatic, as we will see.

6.4.1 Sustainable Forest Management

A term which will recur throughout the following discussion of international forest initiatives, is **sustainable forest management** (SFM). Definitions of what this means vary widely, at some times due to specific field circumstances, at others as a result of the particular purpose it user believes a given forest should be put.

Older graduates of forestry faculties will recall the time when sustainable forest management simply meant the maintenance of the flow of a specific set of goods and services – usually heavily focused on commercial species logs – into perpetuity. As the involvement of the international environmental movement in forests developed through the 1970s and 1980s, leading up to the Rio Earth Summit, this basic materialist line was challenged, often to the point where some groups suggested that the only acceptable definition of sustainable forest management is one which calls for retention of all the assets and qualities of a given ecosystem. In practical terms, this would amount to something close to complete forest protection – a use which of course will be entirely appropriate for some areas of particularly valuable biodiversity and other environmental services, but which would preclude most uses that humans make of most forests they have access to.

Not surprisingly, at the Rio Earth summit, and in its aftermath, the subject of sustainable forest management was widely discussed, and heated arguments arose as to what it meant, or should mean. It became generally accepted that a definition which embodied the intent of the basic definition of sustainable development in the Brundtland Report (cited earlier in this chapter) was required:

> Sustainable development is development that meets the needs of the present, without compromising the ability of future generations to meet their own needs.

A definition of the present day understanding of sustainable forest management was developed by the Ministerial Conference on the Protection of Forests in Europe, and has since been adopted by the United Nation Food and Agriculture Organization. It reads as follows:

> The stewardship and use of forests and forest lands in a way, and at a rate, that maintains their biodiversity, productivity, regeneration capacity, vitality and their potential to fulfill, now and in the future, relevant ecological, economic and social functions, at local, national, and global levels, and that does not cause damage to other ecosystems.

The United Nations Forum on Forests (UNFF) has recognized the difficulty in gaining agreement on a universal definition for sustainable forest management, and has instead identified seven elements which form part of it: (a) extent of forest resources, (b) forest biological diversity, (c) forest health and vitality, (d) productive functions of forest resources, (e) protective functions of forest resources, (f) socio-economic functions of forests, and (g) the legal, policy and institutional framework.

We will not attempt a specific definition of our own in this book. Firstly, because sustainable forest management is a moving target; at any given situation and time it reflects society's values assigned to forest functions. These values change over time and therefore, the definition of forest sustainability also changes. Secondly, the debate has been somewhat subsumed by the advent of forest certification systems that were discussed in Chapter 4 of this book. Certification standards provide a detailed set of criteria and indicators for sustainability which express in operational terms what sustainability means in a specific forest, environmental, economic and social setting at hand: this seems to be the most rational and practical approach.

For our purposes, developing an acceptable and sustainable use for a given forest resource is taken to depend upon the existence of a reasonable consensus amongst all stakeholders that the forest is being used in an appropriate way, and that this requires that the forest will be stabilized, in terms of its biological condition, for an extended period of time. This, in turn, would require that a considered assessment of what forests will be retained into the long term has been made by governments and other stakeholders; and that the uses to which this forest resource will be put will be shared appropriately and then managed carefully, to produce the agreed balance of economic, environmental, social and cultural values.

6.4.2 The Tropical Forestry Action Plan

In the early 1980s, it was becoming increasingly evident to many in the international forests constituency that deforestation, especially in the tropical rainforests, was accelerating. In 1982, an experts meeting was convened by the United Nations Environment Programme (UNEP), the United Nations Educational, Scientific and Cultural organization (UNESCO) and the Food and Agriculture Organization of the UN (FAO). This group recognized that a major international effort would be needed to effectively address this issue, and recommended a multilateral response be implemented as soon as possible.

At around the same time, a major international environmental NGO, the World Resources Institute (WRI) had convened an international task force to devise and implement a programme aimed at reversing rainforest destruction. The United Nations Development Programme (UNDP) and the World Bank became involved in this process. By 1987, the two initiatives had merged, and the resulting Tropical Forestry Action Plan was launched. A more detailed discussion of the history and development of the TFAP and indeed most of the other multilateral initiatives outlined in this section can be found in Humphreys (2006). The TFAP had a brief to support programmes in:

- Forestry in land use
- Forest based industrial development

6.4 A Brief Look at Multilateral Involvement in Forests

- Fuelwood and energy
- Conservation of tropical forest ecosystems
- An action programme on forest institutions

A TFAP Forest Advisors Group was established, with sector experts from the large multilateral banks, UN agencies, participating governments overseas development assistance departments, to provide input into the design and management of the programme, and it was this group which persuaded the managing agencies to translate the objectives of the TFAP into a series of National Forestry Action Programmes (NFAPs), to be triggered by requests from the relevant forested countries.

By this time, the TFAP had three major sponsoring groups – FAO, the World Bank, and the United Nations Development Programme. A large number of developed countries and the multilateral development banks were on the list of financial supporters, and a group of influential NGOs were involved with the official agencies in governance arrangements for the programme.

Before the end of the 1980s, trouble was already brewing for the Tropical Forestry Action *Programme*[6]. In the early1990s, a number of reviews were initiated, Lohman and Colchester (1990) most of which converged on the finding that whatever the TFAP had achieved, it had not produced any slowing in deforestation, which was its basic objective. Nor had it managed to reconcile national interests with the concerns of the international forestry constituency about what was happening in the tropical forests. The NGO groups also complained that local interests had been neglected in the forest countries when formulating and implementing the NFAP programmes. WRM, in its review, ventured the opinion that the TFAP had actually made things worse in forest countries, by facilitating donor funding of projects which were unsustainable in the forests.

Sizer (1994) has pointed out that another recommendation made in the review process was to create an independent consultative mechanism for the program with broad participation and sponsorship. However, FAO refused to expand the governance of the program, against the recommendations of other major participants. Instead, a consultative group was created within FAO, similar in name only to the original proposal. This move further reduced the credibility of FAO as an effective TFAP coordinator and alienated the other three original co-sponsors.

Tensions continued to escalate as the Chairman of the US Senate Committee of Foreign Relations sent a message to the President of the World Bank, urging the Bank to suspend financing of TFAP forest projects, pending further review. By this time, the Bank was in a mood to comply with this request, having had some major disagreements itself with FAO as to the management of some specific programmes within the TFAP. A G7 meeting in 1990 recommended that the TFAP needed strengthening and major reform, and in particular needed to give much stronger emphasis to biodiversity and forest conservation in its work. Some recipient countries

---

[6]At some point, the name was altered from *Plan* to *Programme*.

involved in the TFAP – Brazil, Indonesia and Malaysia – raised sovereignty issues in relation to the manner in which the programme was being designed and implemented. Shortly thereafter, the World Wildlife Fund withdrew its support for and participation in the TFAP, and WRI itself – one of the original advocates of the programme – did likewise. To all intents and purposes, the World Bank was also no longer a participant in the TFAP.

The Forests Advisors Group[7] also withdrew its support, making direct recommendation to the FAO Council that it should not proceed with the TFAP, claiming that the programme had achieved "too little, too late" and should no longer be allowed to distract from efforts being made to start more fruitful undertakings in the sector. The Group threw its support behind the establishment of a consultative process known as the World Commission on Forests and Sustainability.

Sizer (1994 op cit) has noted that the TFAP concept was flawed from the beginning:

> TFAP's problems can be traced partly to the program's inception and launch: implemented as a sectoral planning exercise, it did not take adequate account of deforestation's root causes. The divergent perspectives of governments posed other obstacles. In general, the South emphasized national sovereignty and development while the North pushed for global environmental management. Donors also invested too little in the national exercises. In addition, TFAP was heralded as the "magic bullet" which would halt tropical deforestation, a target which it could clearly never achieve.

In the manner of these things, the TFAP was never officially declared dead, and indeed some aspects of it live on in the FAO's National Forests Programme. These have been accepted as one of the policy tools to promoting sustainable forest management and a large number of country-driven processes are under way to plan and implement such programmes. Donor funding is provided to support these processes in developing countries.

The TFAP experience foreshadowed some political developments in the international forests constituency: the burgeoning role in multilateral programmes of environmental NGOs; the emergence of the sovereignty issue in the multilateral programme in forests; and the building conflict between a production oriented approach to forests that was embodied in the approach of some of the official international agencies involved, and the more political and polemical focus on large, charismatic issues to do with forest conservation, biodiversity protection and local community participation that were emerging elsewhere in the international forests constituency. Since 2002 the focus on poverty reduction has gradually brought attention back to the productive role of forests within the broad framework of sustainable forest management duly recognising the other values of forests.

---

[7]An informal group of principal forest sector advisors from agencies such as the World Bank and other multilateral development banks, the European Community, and some large bilateral development assistance agencies.

6.4.3 The International Tropical Timber Organization

The ITTO was established at around the same time as the TFAP, but managed to keep a much lower political profile – and to an extent continues to do so. The organization's origins can be traced back to a meeting of the United Nations Conference on Trade and Development in 1976, which produced the first International Tropical Timber Agreement. The basis of the idea at the time was that on the one hand, considerable deforestation was occurring in many tropical forest countries, while on the other hand, trade in tropical timber was an important element in economic development in many of these same countries. As it is put in the history of ITTO on the ITTO official website, the reconciliation of these two observations is the ITTO's raison d'etre but it is a complex challenge for a commodity organization, because traditionally these have not addressed environmental conservation, as called for under the International Tropical Timber Agreement.

Thus, despite its trade focused origins, ITTO is not a conventional commodity agreement. The negotiators of the first agreement under which the ITTO was established under UN auspices recognized, early in the piece, that the forest conservation issue was going to be at least as important as trade, and so this is reflected in the objectives of the agreement. Essentially, the projects and programmes the ITTO finances from funds it receives from its donor national members are aimed at the promotion of a trade in tropical timber and timber products from legal and sustainably managed sources.

Most published criticism of the ITTO seems to have been written in the 1990s, with relatively little follow-up to the present time: Gale (1998) argues that the ITTO has bolstered a blocking alliance between the timber industry and producing- and consuming-country governments, which favours developmentalist interests and ideas. He asserts that ITTO has permitted environmentalists to voice their concerns but not to negotiate them.

As will be evident from discussion of the nature of ITTO in Chapter 4, it is unrealistic to expect the ITTO to successfully address deforestation, which is primarily driven by factors outside forests. Timber harvesting is much less of a factor in deforestation than clearing of forests for other purposes. Moreover, in the tropics, only around 5% of that volume of timber is exported. However, it would certainly be the case for Indonesia – especially during the high deforestation period in the 1990s – that the value of trees removed from the forests, and the value added to them in processing prior to export, was in fact a major element of the economy of that period.

6.4.4 The Intergovernmental Dialogue on Forests

In 1992, the Rio Earth Summit adopted the Non-legally Binding Authoritative Statement of Principles for a Global Consensus on the Management, Conservation and Sustainable Management of all Types of Forests, known as the Forest Principles, and included a chapter (Chapter 11) on combating deforestation into the Agenda 21 document, a global plan of action on the environment and economic development adopted by 178 governments at the Rio conference. The Forest Principles offered a useful framework but this achievement fell short of original expectations to reach a legally binding agreement at the Summit.

In response to this major set of international resolutions, the United Nations in 1995 established the Intergovernmental Panel on Forests (IPF) which had the intention of implementing the Forest Principles and Chapter 11 from Agenda 21. This was followed by the Intergovernmental Forum on Forests. From 1995 to 2000, a number of intergovernmental meetings, technical sessions and other activities were convened, around subjects such as international cooperation and assistance for technology transfer, development of criteria and indicators for sustainable management, deforestation, trade and others. The result was a menu of 270 proposals for action for the promotion of sustainable management and conservation of forests.

The full implementation of Agenda 21, the Programme for Further Implementation of Agenda 21 and the Commitments to the Forest Principles, were strongly reaffirmed at the World Summit on Sustainable Development (WSSD) held in Johannesburg, South Africa from 26 August to 4 September 2002.

In 2000, the IFF was wound up, and the United Nations Economic and Social Council replaced it with the United Nations Forum on Forests, which had as its major objective the promotion of the management, conservation and sustainable development of all types of forests and to strengthen long-term political commitments to this end. The principal function of the UNFF was to provide a policy forum for issues related to forests, to facilitate implementation of forest-related agreements and foster a common understanding on sustainable forest management; to enhance international cooperation and to monitor the progress towards agreed objectives and targets. UNFF was also assigned to strengthen political commitment to the management, conservation and sustainable development of all types of forests.

At its sixth session, in 2006, the UNFF announced agreement on four global objectives on forests, to:

- Reverse the loss of forest cover worldwide through sustainable forest management (SFM), including protection, restoration, afforestation and reforestation, and increase efforts to prevent forest degradation
- Enhance forest-based economic, social and environmental benefits, including by improving the livelihoods of forest-dependent people
- Increase significantly the area of sustainably managed forests, including protected forests, and increase the proportion of forest products derived from sustainably managed forests
- Reverse the decline in official development assistance for sustainable forest management and mobilize significantly-increased new and additional financial resources from all sources for the implementation of sustainable forest management

As noted in Chapter 4, on December 17, 2007, the UN General Assembly adopted the Non-Legally Binding Instrument on All Types of Forests negotiated by the UNFF earlier that year.

6.4 A Brief Look at Multilateral Involvement in Forests

The bland, official descriptions of the proceedings and results of the intergovernmental forest dialogue create an impression that much has been considered – which is undoubtedly true – and that much has been achieved – which is not. Throughout the entire duration of these processes, a great deal of frustration and disappointment has been built up: although representation for various stakeholder group representatives was provided many groups have felt excluded and ignored. More recently, UNFF has taken action to improve multi-stakeholder dialogues, and some stakeholders have become better organized to express their views as interest groups.

It was clear from fairly early on that certain participants in the process (for example, the Government of Canada, and some member states of the European Union) had very specific agendas, such as the creation of a global forest convention, along the lines of the Convention on Biodiversity which was signed into by 150 countries at the Rio Earth Summit in 1992. It was also clear that some other participants (e.g. the United States, Brazil) seemed at times to attend the sessions mainly to ensure that such a convention did not emerge. Like-minded states are still exploring common ground for the establishment of a legally binding instrument for forests, as the current non-binding agreement is feared to be too weak to lead to desired change in behaviour among forest stakeholders.

The essential nature of the intergovernmental dialogue was, obviously, dialogue rather than negotiation of difficult international agreements, or implementation of significant field activities. Much of the language in the terms of reference for the process, and its outcomes, is written in the same aspirational terms of future intent that were applied years earlier (in the expressions of resolutions from the Rio Earth Summit itself, for example), indicating that little progress has been made in the intervening period.

In a paper (Reischl 2007) which examines the role of the European Union in the intergovernmental forest negotiations – specifically the failure to produce a forests convention – the author argues that although this might be interpreted as a failure of international environmental governance, the negotiation process itself helped to generate important norms such as sustainable forest management. We must admit we retain some doubts as to what the intergovernmental dialogue under the UNFF has added to the intense and longstanding debate on forests sustainability.

Dimitrov (2005) applies a political science perspective to an assessment of the intergovernmental dialogue on forests, and is much less impressed with the efficacy of the process. He notes that substantive international negotiations on vital issues such as deforestation have repeatedly failed to produce agreement, but that, instead of simply winding up such encounters, at some point governments seemed to decide instead to create the UNFF, which he refers to as "a hollow entity deliberately deprived of decision-making powers". In view of Dimitrov's somewhat mordant view of the intent of the UNFF, his conclusion appears to be that an agency such as the UNFF, given the intentions of some major government participants in the process, cannot be accused of having failed to deliver policy, but rather should be seen as having succeeded in preventing the delivery of policy.

The UNFF is by no means unique in this outcome. No international body can be stronger than its member states. The weakness of the UNFF process results from the desire of some of its member governments to keep it weak, due to their lack of willingness to submit to external regulation in the use of a key natural resource. Both sovereignty and economic interests are at play here: The dual role of forests to provide global public goods and private benefits is difficult to reconcile in an international regime.

6.4.5 The Forest Law Enforcement and Governance Initiatives

A more recent initiative is the FLEG partnership, which links forest developing countries, multilateral development banks, bilateral donor agencies, NGOs, industry partners and other interest groups into a series of regional Ministerial level consultations and follow-up activities on the ground. These are aimed at building political commitment as well as in-country capacity to address illegal logging. Unfortunately, regional processes have achieved only limited practical improvement in country-level forest governance systems, which illustrates yet another example of the difficulty of converting political declarations into practical action.

In 2003 the European Union established an initiative to address the trade aspects of illegal logging, the Action Plan for Forest Law Enforcement, Governance and Trade (FLEGT). The plan combines measures to be taken in both consumer and producer countries, aimed at eliminating illegal timber from trade with the EU. The FLEGT actions have focused on the establishment of bilateral voluntary partnership agreements between the EU and individual trading partners with the purpose of ensuring that wood and wood-based products that are imported from these countries are legal.

In 2008 the US Government modified its Lacey Act which now criminalizes importation of products which have been produced in violation of the US or foreign laws. The EU is preparing similar measures to prevent illegal timber to enter the Community market be it from the member states or imported from abroad (see Chapter 4).

6.5 Forest Policy in the World Bank: Ideas vs Ideologies

The following discussion of forest policy development in the World Bank is not intended to focus attention unduly on that organization, which is just one of many which has had a role in the international forests constituency, but to use the history of forest policy development and implementation in that organization as a case study of how debate on an issue such as this can become dysfunctional.

The Bank, like its regional development bank counterparts – The Asian Development Bank, The African Development Bank, The Inter-American Development Bank, The European Bank for Reconstruction and Development – and many of the multilateral agencies of the United Nations – is a creature of intergovernmental processes, and it is not surprising that in a given sectoral area, such as forestry, what happens inside the

Bank reflects the developments in the international processes going on more broadly. Many of the issues and problems the Bank confronts in its engagement in forests in its client countries are also shared, to a greater or lesser extent, with many of the bilateral donor agencies operating in the sector. Parallels with some of the trends, developments controversies and dysfunctionalities noted above will therefore be seen in some form within the Bank and its equivalent agencies. Ideological and political developments which first influence the Bank will often come to rest in other similar agencies, and lessons which the Bank has learned (or not learned, as the case may be) from such developments can usually be applied more broadly across the international forests constituency.

6.5.1 The "Chilling Effect" of Bank Forests Sector Policy

Like most development assistance agencies, with their multiple sectoral interests and agendas, the Bank is a large and unwieldy organization, characterized by both a steely determination on the part of senior managers to avoid political or reputational risk in lending and policy operations, and by the existence of deep factionalism within the institution as to models of development, and on which sectors or approaches are likely to produce the best outcomes.

This is hardly surprising for an organization which is at once a powerful presence in international development, but is also highly vulnerable to criticism (valid and otherwise) on a wide front of issues, because of its fragmented system of governance[8] in which the political imperatives of member countries can be manipulated by advocacy groups spread throughout the world. Many of the Washington-based social and environmental advocacy groups – and similar groups in other developed world capitals – most involved in this process have often seemed to many observers to exist mainly for this purpose.

In this situation, the matter of what the Bank should do about forests has historically been fraught and difficult: controversies about the treatment of indigenous groups, how local communities should be involved in the development process, what protection biodiversity and other environmental goods should be afforded (and by whom) all play out in the forest environment. In other words, any involvement in this sector can very rapidly become a hotbed of those political and reputational risks so unloved by managers.

Actual investments by the Bank in the forest sector since the 1980s have comprised somewhere between just 1% and 3% of total Bank lending (depending on how the calculation is done, and in what year), so for most operational managers

---

[8]The day-to-day business of the Bank is administered through a hierarchy of appointed managers, from the President of the Bank down. However, matters of Bank policy, and also final approval of Bank lending and other operations, is determined by the Board of Executive Directors. Executive Directors are appointed, usually by the finance ministries of member governments (some Directors represent multiple countries, others large single member countries).

the need to invest in forests is hardly likely to become an institutional survival issue: On the other hand, as sensitivity to environmental and social issues in forests has grown, Bank involvement in any forest situation which turns bad – or at least is successfully portrayed as being so by groups who disapprove of what has been done – can pose disproportionate dangers and certainly very time-consuming problems to managers. Not surprisingly, therefore, there has historically been a tendency for many middle and senior managers in the Bank to tacitly avoid large commitments to the sector. Some have on occasions even sought to make a virtue of this by suggesting that their reticence to invest in forests is driven by a concern that natural forests should be left alone, and not interfered with: an environmental revelation hitherto unsuspected in the individuals involved, and certainly not one that has led to any evident reduction in pressure on these forests in the real world.

This situation was largely precipitated by the release of the first comprehensive forest policy document produced by the Bank, shortly before the Rio Earth Summit (World Bank, 1991). The imminence of the Earth Summit, and the damage done to the Bank's reputation by some high-profile environmental problems in Bank projects implemented in the 1980s, guaranteed that every move Bank staff made in preparation of this policy was watched carefully by forest protection advocacy groups, and a number of aggressive and highly public confrontations took place during the preparation period[9].

The most highly charged words in the policy document which was produced in this tense climate occur on page 64:

> Specifically, the Bank Group will not under any circumstances finance commercial logging in primary tropical moist forests.

This wording, and much of what surrounds it in the third and final chapter, *The Role of the World Bank*, of the document, bears little relationship to what was presented in the preceding Chapter 1, *Challenges for the Forest Sector* and Chapter 2, *Strategies for Forest Development*. This is hardly surprising: much of what appears in the first two chapters was prepared by the original staff team working on the document, and the final chapter was written by others, brought in by senior management to bring the document more into line with what a number of the advocacy groups who had placed great political pressure on Bank management, through the Board of Executive Directors, and in public campaigns, wanted.

Readers may wonder why these few words became such a traumatic issue within the Bank, especially when it is observed that at the time, there were no Bank projects under consideration that *would* have involved direct Bank lending

---

[9]Readers can find an historical outline of the 1991 forest policy process in Wade 1997, although it must be noted here that much of what transpired has not been publicly released. Depending on who is consulted on this period, the performance of Bank senior managers in attempting to defuse this situation has been variously assessed as barely adequate, ranging down to something which has been described by a senior Bank official as "a very long way from Bank senior management's finest hour".

support for commercial logging operations in natural forests[10] (and there would have been few commercial logging operations that would have needed or wanted Bank lending support). However, the reality in the Bank, following the release of this policy, was that the policy very quickly became interpreted by Bank managers and staff much more broadly, in effect to mean that support for any activity that bore any relationship to logging operations in tropical forests was off-limits or, in Bank parlance that was common at the time, was "radioactive". And of course this was exactly the result that those who had pushed hard for this wording to be incorporated into the policy intended.

Opponents outside the Bank of this approach included many of the bilateral development agencies, forest agencies and researchers in developed and developing countries, and a number of large international environmental NGOs who knew the realities of operating in the field in large forested countries. Most saw the new policy as supine and defeatist, and many argued that excluding the Bank from anything to do with logging operations would prevent the organization from taking any significant role in promoting sustainable forest management (of which logging is an integral part), and would certainly not "save" any tropical forest from destruction[11].

However, despite attempts at the time within the Bank to draft operational directives that made the limits of the restriction clear, and representations to the Bank from various interest groups (including from some Bank Executive Directors from large forested countries), the reluctance to engage in natural forest management in the tropics induced by the policy wording prevailed for more than a decade. The policy constraint even had impacts on Bank forest programmes outside the tropics in some cases, even though it did not formally apply to Bank activities outside the tropical regions.

6.5.2 The New World Bank Forests Sector Strategy and Policy

By the mid-1990s, an increasing number of Executive Directors and others had begun to realize that the policy was in fact constraining and counter-productive. This re-think was stimulated by a number of forthright reports from the independent Operations Evaluation Department (OED) of the Bank, which at the time had the role of assessing whether Bank projects, and Bank procedures (including policy directives) were achieving their stated goals. The OED had concluded that the effect of the restriction in the 1991 forests sector policy had gone well beyond a specific exclusion on direct support in Bank lending for commercial logging

---

[10]The Bank's relatively small private sector investment arm, the International Finance Corporation, may well have been considering such projects at the time, but the reality is that the Bank's Board of Directors would not have placed great store on this when deliberating on Bank-wide policy matters.

[11]A prediction that was certainly borne out in deforestation figures in major Bank client countries with extensive forests in the ensuing decade (see following discussion of OED findings, and the foregoing discussion of the international dialogue on forest sustainability).

operations. OED argued that the 1991 policy had in fact become what it termed a "chilling effect" on Bank engagement in the forests sector generally, skewing what financing did go to the sector almost exclusively towards purely conservation activities, or plantation projects (obviously, not ones that involved removal of native forests first), and away from natural forest management, even though this – with all its difficulties – was the only alternative to unsustainable forest exploitation at any scale in the rainforest countries that had any chance of success. OED also noted that the Bank's forests policy had not achieved any discernible downward impact on rates of deforestation in tropical forest countries.

The Board of the Bank expressed an interest to Bank management in re-visiting the 1991 forest policy, and an extensive consultative and analytical program to review it was launched. Predictably, as soon as news of this review became public, a broad swathe of interest groups declared their positions on the issues involved, and it soon became clear that the focus of reaction from a faction of the of environmental advocacy groups, and some social action groups as well, was going to be that, whatever else the Bank came up with as new policy guidelines, retention of the tropical forest logging exclusion in the 1991 policy was going to be the *sine qua non* of any constructive engagement on this subject by these groups.

The extensive (and expensive) processes and activities undertaken in producing a new Bank forest policy, and the background sectoral analyses that supported this are outlined in a strategy document supporting the policy paper, published by the Bank (World Bank 2004). The Bank's new forest sector strategy was built upon three broad objectives for Bank forest sector engagement:

- Harnessing the potential of forests to reduce poverty
- Integrating forests into sustainable economic development
- Protecting vital local and global nvironmental services and values

Among a list of changes in practice and approach given in the forest sector strategy that would be needed to bring about this new, broader focus of investment in this area by the Bank, two are of particular significance in this present context: First, the strategy document acknowledged that the Bank (and by implication, some other large development agencies) had not been particularly successful in linking forest sector concerns into the broad economy-wide policy and institutional reforms that drive economic adjustment programs in developing countries, even though the linkages between these programs, and impacts on forests and other environmental and natural resource assets, was known in many cases to be strong.

Second, the document accepted the fact that the restriction on effective Bank engagement in sustainable forest management in tropical forests embodied in the 1991 document had indeed had a chilling effect on Bank engagement in the sector, as had been argued by the Bank's Operations Evaluation Department since 1999, and that this element of policy needed to change.

When the forest policy was finally approved, in 2002, it became one of what were then known as Operational Policies. These had the role of safeguards over Bank involvement in sensitive and controversial areas – there were about ten of them, covering subjects such as environmental assessment, conditions and rights

6.5 Forest Policy in the World Bank: Ideas vs Ideologies 135

for involvement of indigenous people likely to be affected by in Bank-financed activities, and for people subject to re-settlement, cultural heritage sites, the forests policy, and so on[12].

The new policy itself, as approved by the Board of Executive Directors in 2002, was known as OP 4.36, and details of its contents can be found on the World Bank's website. Perusal of this document will reveal that the specific exclusion on Bank support for commercial logging in tropical rainforest has been replaced by a provision that the Bank should support activities in forests that would lead to acceptable sustainable forest management, with appropriate attention to protection of sites that should be conserved, and the document specified some criteria and indicators to guide Bank staff on how to assess whether the required conditions for involvement and investment have been met in any given case, and how progress is to be monitored throughout the investment period.

The reader will probably have divined, at this stage, that the pathway to finalization of the new provision on logging in tropical rainforest was by no means free of controversy. One of the main protagonists of the argument that no change to the original wording in the 1991 Forest policy on logging in tropical forests would be acceptable was the US based Conservation International (henceforth CI), joined at various times by the Environmental Defense Fund, the Rainforest Action Network and a number of others. Later in this chapter, we will discuss the particular case of CI's position, since it provides an unusual insight into the extent that ideology can overwhelm common sense, when the charismatic tropical rainforests are involved.

These groups argued, before the Bank President and other senior managers on a number of occasions, that not only was any change to the policy wording completely unacceptable, but that the Bank should not in any way endorse the idea, in policy or in its sector strategy, of sustainable forest management as one means to retaining forest cover – in the tropics as elsewhere. CI proposed a specific approach to native forests – as we will outline later in this chapter – and argued that the Bank must accept that it was a preferable replacement to sustainable forest management, in all cases.

Some days before the new forest policy was to go before the Board of Executive Directors in late 2002, for consideration and, ultimately, approval, strong external political pressure was brought to bear on the President of the Bank – Jim Wolfensohn – in effect demanding that the Bank withdraw this new policy from the Board process, and suggesting pointedly that the President's own environmental legacy in the Bank was under severe threat if this did not occur. A strong rumour ran around Washington in the ensuing days that Wolfensohn had caved in to this pressure, and that the position CI and others had been arguing had prevailed – but unanimous approval of the policy by the Board, in a meeting chaired by the President himself just a couple of days later, revealed that this was emphatically not the case.

---

[12]Considerable amalgamation and integration of these policies into broader Bank procedures has occurred since this time, and the details of how policy now works in the Bank can be found on the Bank website.

This series of events reveals the power that ideology can attain, relative to pragmatism and the art of the possible, in certain elements of the international forests constituency. The idea that an organization such as the Bank could simply abandon sustainable forest management as an important plank in the broad project of retaining and protecting as much forest globally as is feasible, was totally unrealistic and unreasonable, yet it was able, in effect, to garnish the support of some high profile political figures, and of many other interest groups, in attempting to place pressure on the Bank to abandon sustainable forest management.

6.5.3 *The Chill Is (Not) Gone*

For better or worse, the Bank's review of policy provided a platform and a focus for debate on some fundamental issues around forestry and development, and interest groups across the spectrum of opinion had been called upon to declare themselves one way or another. Two developments that have occurred since Board approval of the new forest sector policy and strategy that are worthy of note:

It became apparent that once the new policy was approved in the Bank, the rancorous and widespread campaigns being waged on websites and in forums around the world on its content virtually ceased, within a short time frame. Many observers were surprised by this. However, Bank staff and members of other groups closely involved in development of the new policy were less so, having seen that the campaigning activity had more to do with attempts by many of the groups involved in it to leverage their own political position and organizational profiles while the issue was prominent in international forests constituency circles, than with a serious attempt to engage in a pragmatic debate about what might and might not work for donor engagement in natural forest management in developing countries. The strategy in organizations which remain opposed to the Bank's new approach shifted back to raising objections to specific forest sector projects under implementation or consideration by the Bank. As will be noted below, the Bank has left itself open to continuing criticism in this area through an incomplete and in some cases ineffective implementation of the new strategy.

Even more seriously, from the viewpoint of genuine progress in establishing sustainable forest management in natural forests of the developing world, the salutary effects that might have been expected to flow from a new and more progressive policy on the Bank's own engagement in natural forests management have proven to be rather more elusive than was hoped, and indeed was projected in the forest sector strategy document that accompanied the new Bank forest policy. It would seem that some staff and much of management in the Bank continue to regard involvement in anything other than the blandest and lowest profile activities in the sector with trepidation, and the Geiger-counters focused on political and reputational risk from the sector still register strong signals within the organization.

Contreras-Hermosilla and Simula (2007) have reviewed progress with implementation of the Bank's new approach to forests, and have found that in the 5 years that had elapsed since release of its new forests sector policy and strategy, while some progress has been made, much remains to be done:

- While the Bank has had some success in extending its engagement in non-tropical forests (especially in the newly emerging economies of the Russian federation, Georgia, Romania and others), its overall level of engagement in forests remains well below targets set out (and approved by the Board) in the forest sector strategy. The involvement in tropical forests in particular remains modest, and still mired in many cases in controversy. The focus on poverty reduction – a key pillar of the forest sector strategy – has been made appropriately in some cases (the authors cite recent projects in Albania, Gabon and Nicaragua as examples), but not in others. The authors are particularly critical of the continued failure of the Bank to allocate sufficient financial resources to carry out the economic and sector analyses that are needed to identify the potential for effective contributions forests can make to poverty alleviation, economic growth and environmental services, even though this was a major subject of discussion in both the public consultative process that was carried out to support design of the new forest strategy, and in Board deliberations on the new strategy.
- The authors acknowledge that the Bank has made significant efforts in some countries to mainstream the elements of the new strategy, outlined in the previous section, by incorporating forest sector reforms into larger economic strategy documents on country assistance, and poverty reduction, and including broader reforms into forest sector investment projects. In some cases the Bank has also taken into account the impact of its broad-based economic reform programs, based on what the Bank terms Development Policy Lending[13], on forests, and made appropriate adjustments accordingly. However, the authors note that this has not been done in all cases where it should have occurred, thus perpetuating the institutional divide which has built up in the Bank between specific environmental and sectoral outcomes, and broad economic adjustment programs and, in the process, missing opportunities to build progress towards the Millenium Development Goals[14] to which the Bank is a signatory.
- In some internal institutional areas, such as collaboration between the Bank's private sector arm (the International Finance Corporation) and the rest of the organization in sectoral investments, and partnerships with outside agencies such as the alliance with the World Wide Fund for Nature, and the multidonor Program on Forests, there has been some progress, but these initiatives remain less coordinated and systematic than they should be. The authors note that

---

[13]This issue will be discussed in Chapter 7 of this book.

[14]The Millennium Development Goals (MDGs) are eight goals to be achieved by 2015 that respond to the world's main development challenges. They are drawn from actions and targets identified in the **Millennium Declaration** that was adopted by 189 nations-and signed by 147 heads of state and governments during the Millenium Development Summit in September 2000. Forests issues are covered under Goal no 7: Ensure environmental sustainability. More details on this initiative can be found on the United Nations MDG website.

implementation of Bank safeguard policies in the areas of environment and natural habitat protection, indigenous people and re-settlement has been uneven. The Bank's Inspection Panel (which investigates complaints about policy breaches in Bank activities reported by outside observers) has recorded some serious problems with appropriate implementation of policy in some cases related to logging in natural tropical forests. Indeed, it can be expected that any new Bank project in this area will continue to be challenged by some well-organized international NGOs which can, through their local networks, engage a local NGO to submit a claim targeted at an inquiry by the Bank's Inspection Panel. This increases the Bank's image risk, leads to costly delays in project processing and chills the management appetite for any interventions in the forest sector.

6.5.4 Problems with Ideology: The Conservation International case

Conservation International is one of the powerful "Beltway" environmental non-governmental organizations based in Washington which have strong political connections. These NGOs are highly skilled in raising financing and they are competing not only for the same sources of funding but also "flagship" projects and initiatives as part of their image building. It has access to large funds through its contacts with corporate figures such as Gordon Moore[15], and it can reach large segments of the public, through its formidable publicity infrastructure, which has used the services of Hollywood stars such as Harrison Ford to promote its messages and, importantly, the CI brand itself.

In 1998, CI challenged the validity of sustainable forest management (SFM) as a means of replacing the very high rates of loss of natural forests that were occurring in many developing countries – especially in tropical rainforests – with systems of use that would allow retention of more forest, and conservation of some of it.

CI's argument came as something of a shock, because sustainable forest management, regardless of its difficulties (which had led to the plethora of discussion and international agonizing over proposed solutions to its effective implementation discussed earlier in this book) was regarded by most involved in the international forests constituency as essential in the mix of approaches needed to address runaway deforestation. Few practitioners argued that application of SFM in the real world of forest-rich developing countries was ever going to be easy. Their view in general was (and remains) that SFM must be made to work: it is a necessary condition to virtually everything else that needs to happen in forests. Most bilateral

---

[15]Moore is the founder of INTEL, the world's largest semiconductor company and a major presence in the development of microprocessors.

6.5 Forest Policy in the World Bank: Ideas vs Ideologies

development agencies, multilateral development banks, and large international environmental organizations such as the World Wide Fund for Nature (WWF), and the International Union for Conservation of Nature (IUCN), had accepted the inevitability of making the concept work, even if they occasionally disagreed on how this should be done, and what would constitute success.

As discussed in Chapter 2 of this book, the main problem with SFM implementation has been (and to a large extent remains) that there was never sufficient funding available from the international forests constituency to compensate potential losers from replacing exploitive, unsustainable logging operations with sustainable alternatives. Moreover, as noted in the previous sub-section, relatively little has been done to advocate to governments the necessity to link forest outcomes to broader economic policy. The net result has been that governments in countries where deforestation has been significant have not generally accepted (to the extent of developing and implementing effective policy) that the country's best long term interests would be served by enforcing a sustainable management regime in its natural forests – or at least, in a significant proportion of that resource. There were, in fact, relatively few cases where the real economic, social and environmental costs of excessive forest loss were even calculated, let alone incorporated into existing markets and policy.

In a policy note published in Science magazine, the CI authors (Bowles et al. 1998) signalled a particular interest in interdicting the application of SFM in the World Bank which was, at the time, beginning to reconsider its organizational approach to investment in forests (see next section):

> ...recent years have seen a growing criticism of SFM itself – and particularly its utility as a conservation strategy .... The next chapter in this debate is currently taking shape as the World Bank.considers whether to lift its 1991 policy that bars investment in logging operations in primary tropical forests. The Bank's deliberations bring a seemingly abstract debate into sharper focus. The questions before the Bank and its many constituencies are simple: Will new investments in logging operations help to curb deforestation? Can the Bank and its partners bring about sustainable forest management in these operations? And, most important, will this lead to conservation? In our view, the answer to these questions is, broadly speaking, no[16].

In subsequent publications (see, for example, an in-house CI report on the subject by Rice et al. (2001)), CI developed this idea, reviewing historical examples of international development assistance – funded attempts to implement SFM in the field (mostly unsuccessful), and citing prior studies of the economics of sustainable forest management (compared to non-sustainable logging alternatives) – most of which showed SFM to be less profitable, in purely financial terms as evaluated using commercial interest rates, to a more rapid, exploitive logging operation.

As will be evident from the discussion earlier in this section of attempts by the international forests constituency to introduce SFM, CI's diagnosis on this matter was, to say the least, no revelation to most members of that constituency: basic discounting at the commercial interest rates prevalent in developing countries,

---

[16]From Bowles et al, SCIENCE 280:1899 (1998). Reprinted with permission from AAAS.

without convincing valuation of the ecosystem services available from a managed forest, made such a result almost inevitable. This is one of the reasons why many in the conservation movement had promoted, or supported, the use of social discount rates, and the valuation of as many of the non-monetized (or at least, non-marketed) goods and services that intact forests provide, as a basis for evaluating SFM alternatives, since this would be more likely to alert governments and other interest groups to the potentially very high long-run costs of exploitive logging that would not show up in purely financial calculations. Given this, it was quite extraordinary to find a purportedly conservation-oriented organization basing its argument that SFM was a total failure on studies of financial returns to logging, using unadjusted market rates of discount.

CI's alternative to attempting to undertake forest conservation within an SFM context is to allow an initial, usually quite heavy, logging operation to extract the bulk of wood value from a given site, and then to utilize conservation grant funds to finance a closure of the forest to further logging after that (sometimes by means of a conservation concession, whereby the rights to future use of the forest are purchased using the same mechanism under which logging concessions are issued in many countries with large natural forests) – the purpose being, in this case, to allow biodiversity and other ecological values to recover in the long term.

The original location where this approach was developed was in the Chimanes Forest, in Bolivia, and much of what was argued by CI and others attracted to this approach subsequently for other forests in other regions was little more than an indiscriminate generalization of this result. In fact, the situation represented in this forest is quite unusual: the wood values of this particular forest are dominated by very large mahogany trees which occur at very wide spacings amongst a broad range of other species and shrubs. Because of the very high international prices which mahogany logs attract (especially old growth mahogany, as opposed to plantation grown material), this in effect means that more than 90% of the total log value of these forests will be held in these few, old trees. In such a case, it is feasible that removal of virtually all the high value trees could be achieved in an initial logging operation over several years, and that this level of wood value would not then recover on these sites for a very long time – if ever. If carefully done, the initial logging operation would not unduly damage the remaining biodiversity on the site.

However, the same process could not be as easily undertaken, for example, in the dipterocarp forests of Asia, or the equivalent rainforests of the Pacific islands west of the Wallace Line, nor much of the African mixed hardwood forests, nor indeed of much of the rainforest in Latin America either. In all of these cases, purely commercial logging operations, if permitted (or tolerated) would seek to remove much more of the total log volume (much less valuable, on a per unit basis, than the mahogany forests) in these forests, and repeated re-entry into the forests after a fairly short interval as trees that were initially too small to log grow to a commercial size would be feasible. In such cases, the temptation to log the forest initially at rates well in excess of those that would be permitted under a genuinely sustainable silvicultural regime, and then to return continuously to that same forest as sufficient new volume grew to commercial size, would be strong, and the result of such a regime would certainly not be a biodiversity enthusiast's dream; indeed, after a fairly short while, it would in all probability not qualify as a forest at all. This pattern of use has in fact been the norm in many forested areas in the tropics in recent decades, and in most cases the end result of a degraded or even non-existent forest holds no threat to the interests of many of the groups involved in this process, since conversion to other land use has often been their objective from the beginning.

Little wonder then, that even some of the staff of CI itself – especially those with field experience in the forests of South East Asia – were quite alarmed at the organization's proposal. CI's belief that the purchase of a conservation concession following an initial logging will be sufficient to eliminate the likelihood of forest degradation fails a basic reality test: at the scale of global deforestation, the amount of funds that would be needed to pay for the establishment of conservation concessions over a significant area of forest would be enormous. To do so in perpetuity, considering that any future government may not feel itself bound to any agreements originally made on this matter, would be even harder. Consider the case of just one country, Indonesia. At the height of logging operations in the rainforest in that country, in the late 1990s, the value of logs being extracted, under operations that were for the most part unsustainable, averaged in excess of $US 4 billion per annum. A significant proportion of that amount would have been needed to compensate interest groups and stakeholders in that country – ranging from the various levels of governments through to local communities, and including the range of forest industries and oil palm and other potential beneficiaries of deforested sites – for there to be a significant possibility that genuinely sustainable operations in those forests could be maintained. What CI never did, in promoting its approach to the point where exclusion of any SFM option was mandatory, was provide a rational case that its approach could work at a global scale, given the realistic constraints on funding available.

Like many environmental groups, CI's basic raison d'etre for its position on this issue was the perfectly honourable and desirable one of promotion of conservation in natural forests, especially in the tropics. Its approach of log-and-then-leave on a conservation concession basis would be suitable in a forest where the sort of operation it calls for would in any event be close to the commercially preferred operation, or where the area of forest intended for protection is sufficiently small, and sufficiently valuable in terms of biodiversity and other non-wood assets, to justify a large grant. However, this in no way represents the generality of situations that present in tropical forests. In that sense, CI's remedy for excessive and destructive deforestation certainly has no more prospect of success at a global scale than the attempts to introduce SFM have had to date – and indeed, as suggested earlier, almost certainly has less.

Had CI been satisfied, when introducing its alternative to SFM, to advance its critique of SFM itself, and of certification and other aspects of the SFM approach, and to then proceed with its own alternative, that would probably have been the end of the matter, in terms of its public profile. However, as noted in the discussion earlier in this chapter on the Bank's new forest policy, CI and its supporters opted instead to launch a campaign to make SFM actually off-limits to the World Bank and, by extension (or perhaps, in the eyes of CI itself, by example) other development agencies. This is a clear case of where the ends that CI intended to achieve were not justified by the means adopted to pursue them: CI's version of these ends were at the time in global terms simply unattainable.

In pursuing this approach to the extent and by the means it did, CI left the realm of ideas, and entered that of ideology. It sought to force others inexorably to the conclusion it had drawn itself; that its log-and-then-leave solution for forest conservation was, in fact, the only one. CI at this point could be described to have replaced conservation with *conservationism* – a doctrinal, advocacy driven approach that existed to provide continuous justification for a chosen ideology, rather than a pragmatic, results-driven program to move out the boundaries where acceptable management of forests to retain them as intact forests could be implemented.

In essence, our argument is that the problem with SFM, as applied (or not applied) in the heavily forested developing countries, has always been a lack of funds to implement it. As we will show in the following chapter in this book, if there are not sufficient funds to transform incentives on the part of those agents who are causing deforestation – or who are failing to arrest it – then deforestation will continue. Of course, the capacity of agencies responsible for forests to implement regulations which call for retaining forests is often weak; the ability of local communities to engage in forest stewardship is often constrained; the political will to arrest deforestation is usually modest at best, and compromised in many cases by corruption. But none of these problems can be effectively addressed in a financial vacuum.

The same lack of funds would prevent the CI approach from working at scale in any situation where deforestation is intense, but the organization's ideological enthusiasm led it to assume that what has worked at a small scale in a very specific situation would therefore work anywhere. Compromise is often seen as the enemy of an ideology; it is believed to play into the hands of the governing group. Therefore SFM – being essentially a compromise on forest use – was seen by CI as a fatally flawed approach.

6.6 Developing Perspectives on Sustaining Forests

It would be reasonable to suggest that concern for resource sustainability and environmental protection should have remained high on the public debate agenda globally – under the impetus of the many large international colloquiums and conferences that have been held on the subject. However, as we observed in Chapter 2, this has not proven to be the case, and the resolutions from these many meetings has not yet been translated into effective action and policy to address the concerns raised. In the world of national reform and policy responses, the speed of adjustment of modern economies, firstly towards an understanding that degradation of environmental systems and constraints on natural resources are beginning to limit conventional

6.6 Developing Perspectives on Sustaining Forests

economic growth, and secondly towards development of policies that are adequate in design and scale to address this problem, remains slow.

If anything, the microcosm of this situation represented in the forests sector has made even less progress. The international dialogues on forests – especially on what should happen to the world's diminishing store of natural forests – has, as we have noted, been less influential on public opinion than the broader global dialogues on environmental sustainability, and the implementation of appropriate policy responses even less effective than in the general case. The international forests constituency has not to date produced a significant and implementable consensus on what needs to be done, and by whom. It is equally apparent that there are not going to be easy, neat solutions; sustainable forest management is an approach that can work where all, or at least the most important of, interest groups support the objective embedded in the concept, but not otherwise.

The observation we make here, in light of this history, is that there is an asymmetry in the capacity of what might be termed advocacy groups to mount toxic campaigns against what they identify as the enemies of the environment, and the ability of large implementing agencies such as the Bank to counter these campaigns. Despite their claims to field experience and successful programmes on forests, many advocacy groups exist primarily to campaign; but neither the Bank, nor other agencies focused on implementation of investment projects, exist simply to counter such tactics. It was quite evident in the policy process described above that many of the more mainstream NGOs working on forests were opposed to the positions of CI and its allies on this matter, but when they considered whether to raise their voices in defence of a more rational approach, they had to consider the risk that by doing so they would themselves become targets of the radical group, and be forced to invest time and resources into defending their own positions, rather than doing what they existed to do: activities in the field that could improve the environment and the situation of people living in it.

The crux of much of the debate over maintaining natural forests is really the issue of how the various interest groups who have some agency over what happens to those forests actually value them and, in cases where that is leading to perverse results, to what needs to be done to alter perceptions and behaviour of those groups. The next chapter will examine the issue of forest value more closely, with a view to gaining some perspective on approaches that have been applied to this in the past, and what others might be tried in the future.

Trade in forest products, and approaches such as certification of timber as having originated from sustainably managed forests are critical elements in the value question, and this aspect of forest valuation was examined in more detail in Chapter 4.

The forest carbon issue has the potential to be transformative of the whole forests sustainability issue: It has been said by more than one member of the international forests constituency that the potential to market avoided deforestation –or, more directly, the suite of ecosystem benefits that would result from this – in the large natural forests of the developing world offers the first real prospect, in the modern era, to convince those interest groups which currently control forest outcomes in those countries that their forests might, in fact, be worth more alive than dead. This is the subject we move to in the next chapter of this book.

References

Bio-law website http://biolaw.blogspot.com/2008/03/ehrlich-simon-bet-update.html. Accessed Mar 2008
Bowles IA, Rice RE, Mittermeier RA, Da Fonesca GAB (1998) Logging and tropical forest conservation. Science 19 June, 280(5371):1899–1900
Contreras-Hermosilla A (2000) The underlying causes of forest decline, Center for International Forest research (CIFOR) Occasional Paper No 30, Bogor, Indonesia
Contreras-Hermosilla A, Simula M (2007) The world bank forest strategy: review of implementation. World Bank, Washington, DC
Cropper M, Griffiths C (1994) The interaction of population growth and environmental quality. Am Econ Rev: Papers Proc 84:250–254
Culas R, Dutta D (2002) The underlying causes of deforestation and the environmental Kuznets curve: a cross-country analysis. Paper submitted to Econometric Society of Australia meeting
Dimitrov R (Nov 2005) Hostage to norms: States, institutions and global forest politics. Global Environ Polit 4(4):1–24
Ehrlich PR (1968). The Population Bomb. Sierra Club-Ballantine Books
Gale FP (1998) The tropical timber trade regime. Palgrave-Macmillan, New York
Grossman Gene M, Krueger Alan B (1994) Economic growth and the environment. National Bureau of Economic Research, Working Paper No. W4634. Available at SSRN: http://ssrn.com/abstract=227961
Humphreys D (2006) Forest politics: The evolution of international cooperation. James and James-Earthscan, London
Kahn H, Brown W, Martel L (1976) The next 200 years: a scenario for America and the world. Williams Morrow, New York
Kuznets S (1955) Economic growth and inequality. Am Econ Rev 45:1–28
Lohman L, Colchester, M (1990) TFAP: what progress?, World Rainforest Movement, Penang
Meadows DH, Meadows DL, Randers J, Behrens WH III (1972) The limits to growth. Earth Island, London
Meadows DH, Randers J, Meadows DL (2004) Limits to growth, the 30 year update. Chelsea Green, White River Junction, NT
Panayotou T (1995) Environmental degradation at different stages of development. In: Ahmed I, Doelman JA (eds) Beyond Rio: the environmental crisis and sustainable livelihoods in the Third World. Macmillan, London
Reischl G (2007) The EU and the UN Forest Negotiations: a case of failed international environmental governance? Paper presented to Marie Curie European Summer School on Earth System Governance, 24 May–6 June 2007. Vrije Universiteit, Amsterdam
Rice RE, Sugal CA, Ratay SM, Fonesca GA (2001) Sustainable forest management: a review of conventional wisdom, Advances in applied biodiversity science, 3. CABS/Conservation International, Washington, DC
Simula M (2008) Financing needs and flows of the implementation of the non-legally binding instrument on all types of forests. PROFOR. World Bank, Washington
Sizer N (1994) Opportunities to save and sustainably use the world's forests through international cooperation. World Resources Institute, Washington, DC
Speth JG (2008) The bridge at the end of the world. Yale University Press, New Haven, CT
Stern NH (2006) Stern review: the economics of climate change. HM Treasury, Government of the United Kingdom, UK

References

Stern DI, Common MS, Barbier EB (1996) The environmental Kuznets curve and sustainable development. World Dev 24(7)

Sunderlin WD, Pokam J (1998) Economic crisis and forest cover change in Cameroon: the rates of migration, crop diversification and gender division of labour. CIFOR Occasional Paper Series, Bogor Indonesia

Suter K (1999) Fair warning? The club of rome re-visited. Australian Broadcasting Commission. Available at http://www.abc.net.au/science/slab/rome/default.htm. Accessed 2009

van Dieren W (ed) (1995) Taking nature into account: a report to the club of Rome. Springer-Verlag, New York

Vincent JR, Razali bin Mohammed Ali and Associates (1997) Environment and development in a resource-rich economy: Malaysia and the New Economic Policy. Harvard Studies in International Development, Harvard University Press, Cambridge, MA

Wade R (1997) Greening the bank: the struggle over the environment 1970–1995. In: Kapus D, Lewis J, Webb R (eds) The World Bank: its first half century. Brookings Institute, Washington, DC

World Bank (2004) Sustaining forests: a development strategy. The World Bank, Washington, DC

World Bank (1992) World Development Report. Washington, DC

Chapter 7
Financing Forests Sustainability from Ecosystem Values

Abstract The chapter begins by suggesting that sustainable management has rarely been achieved in forests in the developing world not because the technologies for sustainable forest management are not viable, nor because resource managers do not recognize or understand the concept of sustainability. Rather, failure can be attributed to the fact that much of what happens in forests is determined by decisions made far from the forests sector, and to the fact that many of the actors in the forest sector (or nearby) are operating under perverse incentives.

Attempts to quantify the non-marketed ecosystem values of forests have encountered major methodological problems, while the impacts of macroeconomic policies on forests vary considerably from one place to another, and there has been little fieldwork on this subject in specific locations. These constraints, added to the observation made in Chapter 6 that development assistance financing has not been adequate to value sustainable management highly in the eyes of the agents of forest change, illustrate the difficulty of implementing sustainable forest management in the recent past.

Climate change, and increasing global awareness of the fragility of natural ecosystems is the potential game-changer here, and the remainder of the chapter explores the issues surrounding the reduced emissions from deforestation and forest degradation (REDD) mechanism which could finance sustainability via carbon emission trading from reduced deforestation, and an alternative approach based on public/private sector financing of forest ecosystems which does not rely on a specific trading mechanism, but does nevertheless apply value to retained forest carbon, among other assets.

We have now reached a point in this book where we need to begin to draw together the large issues of economic growth and change, forest sustainability, and incentives, and review how these bear on the question of deforestation. Our aim here is to try to develop ideas of what might work to address that large question proactively, rather than to join in the formulation of ideological positions on the question which, as we will see, happens in this part of the policy universe for forests, as elsewhere.

7.1 The Failure of Forests Sustainability: A Question of Perceived Value

In Oscar Wilde's novel *The Picture of Dorian Gray* a character accuses certain people of "knowing the price of everything, and the value of nothing". The target of this sentiment has since variously been directed at economists, politicians, accountants, and even cynics. Elements of this feeling will attach to some of what follows: those who value rainforests highly, for all sorts of intrinsic and intangible reasons, become extremely frustrated with those who insist that these forests must be able to "pay their way" in practical ways – usually involving the exchange of money.

One thing we believe is implied in the following discussion is that the intense debate which has been happening in recent decades over definitions and technical approaches to sustainable forest management in developing countries, and especially in the rich tropical rainforests, has been largely futile; a modern day equivalent of the mediaeval theological debates as to how many angels can dance on the head of a pin. The fact of the matter is that forests sustainability in the tropics, by any definition ranging from the prosaic attainment of log supply in perpetuity through to the most complex inclusion of all forest ecosystem values, has rarely been attained at any scale. A central question we have been trying to address in this book is: what has been missing in our attempts to achieve sustainability so far?

Our underlying view on the sustainability question was introduced in Chapter 2, and has been emerging throughout this book: Sustainable management of natural forests has failed in many parts of the world not because the technologies of managing forests in this way are non-viable, nor because the goals of management systems in place do not recognize or understand or support all elements of sustainability. It has failed because of two central issues of policy and incentive in the political economy of this sector:

First, many of the economic and social policies influencing forests and forest dependent people are initiated a long way from the forest sector itself and can only effectively be manipulated by mechanisms that operate well outside the sector. Government policies designed to promote the development of exports based on a large non-renewable natural resource (oil is the classic example), or to intensify agriculture, or develop domestic manufacturing behind tariff and non-tariff protection policies, are cases in point. Development policies have often been adopted for reasons that have little or nothing to do with concerns about forests, but which can have major impacts (which can be good or bad) on them. Another case is where changes in external trade conditions for a given country create an incentive for rapid expansion of agriculture[1]. This can also have good or bad impacts on forests, although historically it seems to have been primarily the latter.

---

[1] Rapid expansion of soya growing in Brazil, or oil palm plantation in Southeast Asia (both discussed in Chapter 5) are examples.

It is likely that the dynamics of such large economic and social changes have been determined at senior levels of economic policymaking, in response to the political agenda of the government, or to forces originating outside the country: Either way, the reality in many countries is that these changes have almost certainly not involved prior consultation with the interest groups most concerned with forest outcomes – in many cases not even with the ministers for forests or for environment themselves. Bringing in these groups after the fact, in an attempt to mitigate adverse impacts on forests resulting from such large scale exogenous changes, is likely to be an ineffective – or at least an inefficient – approach. Changing perceptions on forest values held by the individuals and groups who are actually responsible for national economic and social policies would be more successful, but this would require very different approaches on the part of forest sector interest groups.

The second reason for the failure of forests sustainability to be mainstreamed rests on a group of incentive issues related to important stakeholders who *are* closely involved in the forests – government forest and environmental agencies, forest owners, private sector operators, local communities, and others – any (or all) of whom have not been convinced of the broad and long term benefits of sustaining forests. These groups will not have made the necessary compromises to their own interests, nor they have been sufficiently driven by public concern over the consequences of not pursuing it to do so.

In this view, we are supported by the observations of the late eminent environmental economist, David Pearce (2001) who was of the view that global deforestation had reached a level where it now imposes risks to ecological resilience and human well-being, and that without an understanding of its causes, effective policy to address it cannot be designed. He goes on to suggest that many governments provide financial incentives to convert forest land – often aimed at rent-seekers – while at the same time many of the ecological functions of the forest remain un-marketed, creating the illusion that since their price is zero, their economic value must also be zero.

Our task in this chapter is to look more closely at questions of forest value – especially that of *whose* value is being considered – and then to review some of the options for closing dichotomies between valuations of forests held by different groups who have some sort of agency in those forests.

7.1.1 *Valuing the Natural Forests: Qualitative Assessments*

In the opening pages of this book, we discussed the charismatic nature of the great rainforests, and the distress that witnessing their destruction creates in members of the public in developed countries, and in many of the members of the international forests constituency. Qualitative or even subjective views of rainforest value have their place, and can certainly be influential on determining what large groups of people think should be happening in rainforests. It is not always possible to derive quantitative measures of rainforest worth and, as we will see later, even when this is done, the figures provided are by no means unanimous or non-controversial. For this reason, we do need to review some of the qualitative views on rainforest value.

A useful summary of the range of ecosystem services that rainforests provide can be found at Annex 2 of the Prince's Rainforest Project Consultative Document (Prince's Rainforests Project, 2008) About 30 categories of service are listed, under headings of provisioning services (foods, fibres, fuel, fresh water, chemicals and pharmaceuticals), regulating services (air, climate, water provision and purification, erosion disease, pests, natural hazards), cultural services (recreation and tourism, ethical values), and supporting services (nutrient and water cycling, primary production). Many of these services are of vital importance to large numbers of people who live in or near rainforests, but some are of much broader consequence: The medicinal value of many rainforest plants has been well-known to communities living in or near these forests for centuries, but it is also the case now that some 25% of modern pharmaceuticals are based on compounds extracted from rainforest plants, and significant elements of the gene stocks which have produced fast-growing crops and other plants originated from rainforests (although under the heading *Quantifying ecosystem values* below in this chapter, we note that while this has undoubtedly been of great value to humanity, we need to bear in mind Professor Pearce's observation that that value has not translated into significant financial rewards for the owners of the forest resources in question as yet).

There are some specific aspects and values of tropical rainforests which potentially have global implications which we have examined in this book – primarily in Chapter 3 – and we note them in point form for reference here:

- *Rainforests, clouds and rainfall.* In Chapter 3, we reviewed the global impacts of rainforests in the tropics on cloud formation, rainfall and planetary cooling.
- *Potential for conflict.* In Chapter 3 we also noted the potential for international conflict over the impacts of declining rainforests: catchment effects on downstream river flows that cross national boundaries; and smoke haze from uncontrolled fires in rainforests that does the same.
- *Cultural values.* Rainforests are also the repository of less tangible but in some respects deeper values to humanity in general: they contain large numbers of sites sacred to many people. Even the biodiversity issue (rainforests contain some 50% of all species of life on Earth), while often couched in scientific terms such as the importance of retaining gene pools and the delicate ecological balance that these complex systems themselves depend upon, also has more fundamental meaning to people who see the rich diversity of life on this planet as something of central importance in and of itself.

7.1.2 Quantifying Ecosystem Values

There have been numerous attempts by ecologists and environmental economists to calculate an economic value for ecosystems, but it has proven to be a controversial and fraught subject for the last couple of decades. There are a number of approaches economists have applied to this task, including avoided costs, replacement cost, factor income, travel costs, hedonic pricing and contingent valuation.

An early attempt to calculate the value of ecosystems in general at a global level by Costanza et al (1997) used prior studies and information to measure the economic value of 17 ecosystem services for 16 biomes. This yielded a range of annual costs of US$16–55 trillion; an average of US$33 trillion per annum. It is clear that forest ecosystems – especially the tropical rainforests – would have constituted a significant proportion of this aggregate valuation.

Not surprisingly, given the enormity of this figure, this estimate received considerable criticism. Bockstael et al (2000) argue that the study values small changes in ecosystems while holding all else constant, and that this constitutes a "failure of additivity". Pearce (1998) argues that the results are inconsistent with the "willingness to pay" criterion, because the figure of US$33 trillion exceeded total measured global GDP at the time, and that the willingness-to-pay measure can only be applied to small, limited valuations.

More recently, in a collection of articles edited by Braat and Ten Brink (2008) there are estimates of the costs of damage to all forest ecosystems (including climate change regulation) at US$1.81–4.14 trillion per annum.

Pearce (op cit 2001) has commented on the results of attempting to build monetary value into various elements of ecosystem services. He concludes that:

- Early optimism about the prospects of genetic material contained within forests with potential for drug development and crop innovation has not been borne out.
- Watershed benefits from intact forests have not been particularly high, but in some cases have been able to offset the benefits of deforestation.
- Reduction of discount rates for pioneer agriculturalists in or near forest areas (via targeted credit subsidies) would be an effective stimulant to sustainable agriculture (in place of the deforesting variety).
- The potential value of forests as a stock of scientific information is not yet known.

Pearce makes two strong points, in this context. The first is that those who place their faith in the achievement of sustainable forestry, in the absence of monetization of presently non-marketed forest ecosystem services will be disappointed: sustainable forestry pays; but unsustainable forestry pays much more. The second is that the carbon storage values of natural forests are going to be of primary importance: those arguing against including forest carbon in permit and offset schemes are in effect removing a major economic argument for conservation of these forests.

This observation – now widely accepted in forest policy making circles – provides us with a convenient segue into the issue of the potential value of stored forest carbon.

7.1.3 Forests and Climate Change

As the reality of climate change itself, and the issue of how to address it, continue to develop in the consciousness of the global public, there is no doubt that the forests carbon factor – especially that arising from the huge mass of carbon stored in natural forests – is a game-changer so far as global forests sustainability is concerned.

It could act upon those perceptions and perverse incentives surrounding the use and abuse of forests that we have discussed throughout this book in a way that has simply not been possible in the past. It could (and we hope will) render the specifics of much of the conflict, debate and fragmentation of effort in the international forests constituency that we have raised in earlier chapters of this book nugatory, or at least marginal. On the other hand, however, it opens up potential for divisions over new issues, and Professor Pearce's warning outlined in the previous paragraph remains relevant: the same sort of single-issue campaigning, and attempts to load specific agendas into general initiatives that has afflicted the dialogue over forests in the past could emerge all over again.

We assume that any reader who has reached this point in this book will most likely have accepted the basic science of the global warming argument, and the human origin of significant increases in emissions of greenhouse gases over the past 100–150 years, and we do not intend to reprise the case for this position here. As far as we are concerned, such doubt as continues to be expressed by the rapidly diminishing group of climate change denialists and sceptics has been comprehensively countered by mainstream analyses and reasoning.

Nor do we intend here to engage further in the global dialogue which has gone on over the past few years on the implications for humanity of the warming trend: we accept the basic argument of the Intergovernmental Panel on Climate Change, and its progenitor, the United Nations Framework Convention on Climate Change, that unless addressed very soon, global warming will cause major global environmental damage with serious – even existential – consequences for humanity.

In short, our view is that, while there remain significant analytical differences within the informed scientific community over the specific mechanisms involved in global warming, there is no serious doubt about the existence of the warming phenomenon, nor its causes, nor its potential environmental seriousness.

In light of this, all we need to do here is to remind readers of some of the basic conclusions which were put forward in the Intergovernmental Panel on Climate Change Fourth Assessment Report (IPCC 2007) on global warming, its likely consequences and options for addressing it. We have summarized the relevant ones below:

- Warming of the climate system is unequivocal, as is now evident from observations of increases in global average air and ocean temperatures, widespread melting of snow and ice and rising global average sea level.
- Observational evidence from all continents and most oceans shows that many natural systems are being affected by regional climate changes, particularly temperature increases.
- There is medium confidence that other effects of regional climate change on natural and human environments are emerging, although many are difficult to discern due to adaptation and non-climatic drivers.
- Global GHG emissions due to human activities have grown since pre-industrial times, with an increase of 70% between 1970 and 2004.
- Global atmospheric concentrations of CO_2, methane (CH_4) and nitrous oxide (N_2O) have increased markedly as a result of human activities since 1750 and

7.1 The Failure of Forests Sustainability: A Question of Perceived Value 153

now far exceed pre-industrial values determined from ice cores spanning many thousands of years.
- Most of the observed increase in global average temperatures since the mid-twentieth century is very likely due to the observed increase in anthropogenic GHG concentrations.
- There is high agreement and much evidence that with current climate change mitigation policies and related sustainable development practices, global GHG emissions will continue to grow over the next few decades. Continued GHG emissions at or above current rates would cause further warming and induce many changes in the global climate system during the twenty-first century that would very likely be larger than those observed during the twentieth century.
- Both bottom-up and top-down studies indicate that there is high agreement and much evidence of substantial economic potential for the mitigation of global GHG emissions over the coming decades that could offset the projected growth of global emissions or reduce emissions below current levels.
- Many options for reducing global GHG emissions through international cooperation exist. There is high agreement and much evidence that notable achievements of the UNFCCC and its Kyoto Protocol are the establishment of a global response to climate change, stimulation of an array of national policies, and the creation of an international carbon market and new institutional mechanisms that may provide the foundation for future mitigation efforts. Progress has also been made in addressing adaptation within the UNFCCC and additional international initiatives have been suggested.
- There is high confidence that neither adaptation nor mitigation alone can avoid all climate change impacts; however, they can complement each other and together can significantly reduce the risks of climate change.

In recent years, more economists have joined the case for action on climate change. The best known and most influential work on the economics of climate change is the Stern Review (Stern 2006), which expanded the scope of discussion of the climate issue beyond the scientific and technical aspects of the subject. It focused on the question of what addressing the climate change issue with substantial amelioration measures might cost, and compared this with the costs of continuation of a business-as-usual scenario. Results from Stern's economic modelling suggest that a failure to act on climate change will generate costs and risks of at least 5% of GDP each year, and inclusion of a wider range of risks and impacts will raise that estimate to 20% of GDP or more. In contrast, Stern suggested the costs of reducing greenhouse gases sufficiently to avoid the worst impacts of climate change could be limited to about 1% of GDP per year: however, in light of new scientific evidence that some phenomena related to global warming appear to be occurring earlier and more intensively than indicated in the IPCC Fourth Assessment, Stern has recently suggested that the effective abatement cost figure might now best be thought of as likely to be nearer 2% of GDP, than the original 1% estimate. In the Review, Stern acknowledged deforestation as an important contributor to rising GHG emissions, and suggested reducing deforestation would be one of the more important pathways to GHG emission reduction.

A recent study undertaken by McKinsey & Company (McKinsey 2009) of the relative costs of abatement of GHG emissions from all available alternatives calculates that to get the world on a pathway to the IPCC target atmospheric concentration of 445–490 parts per million (ppm)[2], a reduction (compared to business-as-usual projections of GHG emissions) in annual GHG emissions of 17 Gt[3] per annum by year 2020 would be needed, and this figure assumes an overshoot reaching a maximum of 510 ppm will occur en route to that figure, even if reductions in human-induced GHG emissions begin immediately.

To place this figure in some perspective, and give some idea of colossal scale of the emissions involved, the US Department of Energy Office of Science (USDE 2005) provides the following estimates of what it would take to reduce emissions of GHG by 1 Gt, under various options, including: replacement of 1,000 conventional coal fired power plants with zero emission alternatives; construction of five hundred 1 GW nuclear power plants; replacement of 1 billion motor vehicles running at 20 mpg with the same number running at 40 mpg.

The carbon content of rainforests is going to be of primary importance in the task of reducing global GHG emissions overall: According to the Intergovernmental Panel on Climate Change (IPCC 2007 op cit), global deforestation is responsible for something in the range of 17–20% of all human-induced greenhouse gas (GHG) emissions. This makes deforestation the second largest cause of these emissions, after power generation and ahead of transport, industry, agriculture, buildings and others.

Some analysts have suggested use of a lower figure for the proportional contribution of forests loss to aggregate human-induced GHG emissions than the IPCC range suggested above; a range of between 12% and 15% has been suggested. This is suggested because continued rapid rises in industry and energy emissions globally in the recent several years, combined with a modest slowing of deforestation rates compared to figures derived on the basis of 1990s and early 2000s data, suggests the contribution of forest loss may have declined. However, we believe there are two factors which may offset these observations:

- First, significant areas of rainforest are situated on peat soils[4], especially in South East Asia. These soils contain huge amounts of carbon – orders of magnitude higher when they are relatively deep than even the very high carbon biomass contained in trees that occupy these sites when rainforests are involved. Deforestation does not necessarily release the carbon from peat soils, but the process of conversion of those lands to agricultural purposes – usually involving draining of the soils – does result in release of large amounts of carbon in the

---

[2] A number of climate scientists – notably James Hansen of NASA – have argued that even this figure will be inadequate to provide reasonable amelioration of some of the worst potential consequences of climate change, and that reductions in GHG concentrations down to the range of 330–380 ppm will need to be attained to achieve this.

[3] A gigatonne is equal to one billion tonnes.

[4] Peat is formed when plant material in marsh or swamp areas is subjected to limited decay under anaerobic conditions from a limited decay process when plant material in marsh or swamp areas.

form of carbon dioxide. It is reasonable to suggest, therefore, that since such conversion is often the fate of deforested areas, removal of that forest will correlate with high losses of carbon through draining of peat soils which follows, in addition to the forest carbon loss incurred.
- Second, some recent research (see Luyssaert et al. 2008; Lewis et al. 2008) suggests that mature tropical rainforests in fact continue to absorb carbon into their biomass, at a rate of approximately 1–2 t per ha per annum. This is contrary to what was previously thought to be the case, which was that forests in this state did not continue to absorb carbon and, in fact actually emitted small amounts of carbon. If this new finding is borne out in larger scale analyses, it would imply that when rainforest is removed, not only is the carbon stored in the biomass of the trees released, but the potential of that forest to continue absorbing carbon in the future is also lost.

On these bases, we therefore suggest that the higher range of estimates of the contribution of deforestation to global anthropogenic GHG emissions presented by IPCC may still represent a more realistic estimate for the present than the lower figures shown earlier. Even if *all* other sources of carbon dioxide emissions were to eliminated, that arising from deforestation at present levels alone would exceed the 450 ppm figure in 25–30 years. In contrast to many of the other major contributors to climate change, it is possible to take short-term action that will reduce emissions and slow the rate of advance on the critical 450 ppm CO2 concentration figure – currently 3 ppm per annum. No new technology is required to buy the world critical time by stopping deforestation, making it a realistic policy objective when it is considered that many other actions to address climate change will take time to deliver tangible results.

Tropical moist rainforests dominate in the discussion of forest carbon and climate change, firstly because they are most at risk of deforestation, on a proportional basis, and secondly because the carbon content of rainforests has – until recently – been believed to be higher, per unit of area, than other types of forest: as was noted in Chapter 3, a recent study by Keith et al. (2009) sheds some doubt on this claim. However, from the viewpoint of prioritizing where to concentrate future efforts to reduce deforestation, it is certainly still the case that the tropical rainforests are most vulnerable to continued loss. One study (Houghton 2005) suggests that tropical deforestation accounts for 96% of global forestry emissions, and that humid tropical forests account for most of this. In aggregate, there is more carbon stored in rainforests than there is in the atmosphere.

A report prepared by The Prince's Rainforests Project (PRP) for presentation to leaders of the G20 nations in London in April 2009 (Prince's Rainforest Project 2009) notes that significantly reducing tropical deforestation could alone reduce global carbon emissions by 5 Gt of GHG per annum: comparing this to the McKinsey target aggregate emissions reduction figure of 17 Gt of GHG per annum cited earlier illustrates the significance of this source of reductions. As we will see later in this chapter, reducing deforestation has the potential to be a (relatively) low cost option for GHG reduction, and this will be an increasingly important consideration as high emitter nations move from commitment to implementation of reduction strategies.

7.2 Stored Forest Carbon: Leading the New Sustainability Paradigm

The objective, or motto, of The Prince's Rainforests Project is simple, yet telling: "making the rainforests worth more alive than dead." It subscribes to the view that intact rainforests now have high option value, which is comprised of presently (or imminently) marketable products – dominated by carbon – and potentially marketable values: maintaining the health of large, multinational waterways; rainfall over wide geographical range; genetic diversity and related values; and so on. However, it also acknowledges that there is a task – indeed a prodigious one – involved in bringing this objective to reality.

7.2.1 Will Reducing Tropical Rainforest Deforestation Be a Cost Effective GHG Strategy?

An accurate answer to this question would require an extensive global study of the deforestation reduction options in the field, and this work has not been done as yet. In the meantime, we can report on some global estimates which have been made of the opportunity costs of reducing deforestation, based on the economic benefits of alternative crops and other activities that would be undertaken the land vacated by the forest lost.

A paper prepared by Grieg-Gran (2006) for the Stern review of the economics of climate change estimates the costs of reducing the rate of global deforestation by half within a decade (from 2005). The study is based on eight countries for which some data were available: these countries together have a total annual rate of deforestation of 6.2 million ha (i.e. close to half the annual deforestation figure of 13 million ha per annum), and the calculation done assumes deforestation is eliminated from these eight countries. Assumptions are made on a country-by-country basis as to whether and how returns to the original timber harvesting to clear the land should be dealt with, and on what distribution of land among high and low value crops will result in the different cases represented. Annual returns per hectare are converted to net present value per hectare at a 10% rate of discount over a 30 year analysis period.

In what the author considers to be the most realistic scenario, the opportunity costs of foregone production amount to US$5 billion per annum. Assuming that the highest value land use occupies all land deforested each year in each country would yield a figure of US$11 billion. Costs of administering the scheme are given as ranging from US$25–$93 million per annum, and since this figure accumulates as

7.2 Stored Forest Carbon: Leading the New Sustainability Paradigm

more area is brought under the scheme, by year 10 the figures for these costs are US$250–930 million.

For comparison with another estimate, these figures would yield a global opportunity cost of reducing deforestation to zero over 10 years of US$11.1 billion per annum. This is close to the figure of US$12.2 billion per annum estimated by Blaser and Robledo (2007) in a study for the UNFCCC.

It is very likely that the average net output figures for crops and other land uses encroaching onto rainforest are now higher than the figure of US$815 which can be derived from the production value figures given in Grieg-Gran. There are two reasons for suggesting this: first, as will be seen from figures cited in Chapter 5 of this book, production and export levels of the higher value land uses in the largest deforesting countries have risen sharply from 2004/2005 to 2008/2009: In Indonesia production of oil palm rose by about 8% per annum over this period, and exports rose by 9% per annum. In Brazil, soy production rose by 5% per annum, and exports by 8% per annum; and beef exports by 7% per annum. Second, the prices of these products on export markets has also risen sharply[5]. China's imports of soy and palm oil have risen by 8% per annum over the 5 year period, and its own production of soy in this same period has actually fallen.

Since the opportunity cost being calculated is a function of the level of output and price, it can be seen that it will probably have moved up significantly since 2005. The rate of deforestation itself has not moved upward dramatically over this same period, which suggests that these high value crops must have occupied a larger aggregate share of deforested land during this period. As long as these trends in production and price continue (failing any relocation of these activities to unforested lands in the high deforesting countries), this will become increasingly the case.

In the Eliasch review (Eliasch et al. 2008), an updating of the Grieg-Gran estimate is made, raising the opportunity costs from US$5 billion–$7 billion per annum. The review provides a number of caveats to this estimate, each suggesting that the figure might be an underestimate. The Eliasch review further estimates that at a global deforestation level, an additional US$4 billion, spread over the first 5 years of the programme, and covering 40 nations participating, would be needed to cover capacity building costs. If it is assumed that base opportunity costs as measured by Grieg-Gran in 2004–2005 had increased by 40% from those years, then the base cost figure would now be in the order of US$15 billion per annum. If the Eliasch figures for capacity building costs are added in, and a generous allowance is made for transactions costs that would be involved for both buyers and sellers, it is reasonable to suggest on the basis of these figures that the total cost of reducing deforestation over a 10 year period should be no more than US$25 billion per annum.

This will sound like a large sum – albeit, in the present era of trillion dollar bailouts for financial institutions during the global economic crisis that was in train at

---

[5] The FAO Trade and Markets website has index figures for oil seed prices that show prices rose by a factor of 3 from 1998–2000 (the index base period) to 2007, before falling back to 2 in 2008, in response to the international financial situation.

the time of writing, not all that large. The key comparison, however, should be with the costs of alternatives for emission reduction. In the Stern review, the overall cost of reducing GHG emissions to the target levels is given as about 1% of global GDP per annum (which would amount to about US$600 billion per annum at present). If deforestation were able to be eliminated in the tropics, and if this represents 20% of all emissions, then on a pro rata basis, this would consume about US$120 billion per annum of that total costs figure. However, as shown above, the studies available show the *actual* costs of bringing reduced deforestation credits to the market in aggregate terms are estimated to be significantly less than this.

In brief, on this basis, the costs of abatement of the 20% or so of emissions from forest loss, avoided deforestation, would be only 4% of the total cost of abatement of emissions according to the Stern formulation.

Another way of analyzing the cost-effectiveness of avoided deforestation is to compare the economic benefit of the carbon value of retained forests with specific net returns from the cash crops and other uses which are made of deforested land. Some notional calculations of this for some of the competing crops outlined in Chapter 5 show that a price for forest carbon (US$45 per ton or more) would be required in those cases to offset the stream of benefits of deforestation for purposes of establishing alternative crops or other land uses on the deforested land, at current commodity prices. This is not an enormous price for carbon; it translates into a little over US$12 per ton of carbon dioxide (the more common way of expressing prices for GHG emissions), and this is by no means high in comparison to the costs of GHG abatement via emission reduction technologies presently available within developed countries.

The commodity prices used to derive these figures are significantly lower than current market prices, for some of the products at least. On the other hand, it is important to recognize that the comparison as calculated above assumes *total loss* of production of the alternative crop when calculating the opportunity cost; the global estimates of opportunity costs of forest carbon cited above share this assumption. But the reality more often is that an alternative crop could be wholly or partly displaced onto other land – sometimes land that might presently be utilized for a lower value crop (in which case the net value of that crop becomes the opportunity cost of locating the previously deforesting commodity there); and sometimes including degraded or otherwise marginal land, presently used for very low value production purposes, if at all (in which case the opportunity cost becomes the costs of rendering that land as favourable for production of the commodity as the forest land it would have occupied)[6].

We acknowledge here that implementation of reductions of deforestation in practice on the basis of carbon values will need go beyond the broad-canvas figures we have presented here, to take many local considerations into account: food

---

[6] There is also the question of who receives the value of logs which would have been harvested from the forest site prior to conversion to the competing commodity: we discuss this issue in the case study below.

security; liquidity of assets; financial and natural risk mitigation; pre-existing forest degradation; the potential of plantations on some sites, and so on.

7.2.2 A Case Study: Oil Palm in Indonesia

The best way to gain some insight as to the practical economic reality of displacing commodities which have to date been replacing forested lands, with the objective of producing a carbon value from the forest retained, is to look at a particular example. The one we have chosen is oil palm, in Indonesia. This case is useful because it represents perhaps the most high valued conversion use to which forested land is being put in tropical rainforest countries; if the carbon option through retaining forests looks reasonable in economic terms in this case, then – in very general terms – it is likely to compare favourably to most other uses for which tropical forest land is currently being converted.

Developments in the palm oil market: Oil palm occupies a substantial area of land in Indonesia, estimated at around 6.1 million hectare in 2006. The industry has grown very rapidly, and international demand for the product seems likely to continue to be strong, since, as a highly traded food staple, this is likely to be less affected by the economic downturn than many other commodities. A study of oil palm prospects by Corley (2008) sees demand for the product doubling on global markets by 2050. Meeting this demand would require establishment of an additional 250,000 hectares of plantation each year until then. On current indications, Indonesia could be expected to provide more than half of that increase; its share of global production has risen sharply in recent years, and at 18.2 million ton (2007) is already about half of the global total.

Developments in the relatively new biofuel market (see discussion in Chapter 5) could drive the rate of demand significantly higher than these figures suggest. According to Corley, sufficient oil palm and other oil crops could be grown on already open land to meet his projected demand for these products for food by 2050, but a continued rise in the use of these products for fuel manufacture would compromise this outlook significantly. This has become an issue at the European Parliament, which has passed directives requiring minimum emission benchmarks to be applied to palm oil imports intended for use in biofuel. The European Union Environment Commissioner has recently expressed doubts about the sustainability of biofuel as presently produced.

A large part of the area of Indonesia planted to oil palm was originally occupied by tropical rainforest, and it is primarily for this reason that the palm oil sector is seen by many as a major driver of deforestation in Indonesia. We have suggested, in Chapter 5, that identifying causality is in fact not straightforward, in this case. Nevertheless, the perception of oil palm as a major deforesting agent is generating a series of intense encounters at the Roundtable on Sustainable Palm Oil (set up in 2001 by large retailers, producers of palm oil, and public interest groups to promote ethical and ecologically responsible production). It has also ignited some public campaigns against the use of oil palm from unsustainable operations (meaning operations which occupy land cleared of tropical rainforest for this purpose, or peat lands, which in general terms contain much more carbon than that stored in rainforests).

Whatever the merits of the public campaigns, oil palm is going to be a major presence on global markets for the foreseeable future. It provides per unit area oil yields ten times greater or more than other vegetable oil producing crops (including soy beans): this could mean that any market intervention to reduce its presence on global markets might simply drive demand higher for other crops which may have even more deleterious impacts on the forest environment. For this reason, if there is a feasible pathway to locate future oil palm plantations on un-forested lands (particularly those which are presently under fairly low productivity usage), then it would be advisable to try this approach first, before measures to drive unsustainable oil palm out of the global market are applied: Oil palm production is not environmentally unsustainable per se; it is unsustainable when it is being located on areas of intact forest – and only then if it the primary cause of that deforestation (see discussion of this in the final section of Chapter 5).

Some investment at the margins to provide an incentive to plant new oil palm estates on open sites (which are informally defined here as land which has not had intact forest cover for a considerable period: the CDM cut off date of 1990 would be a reasonable criterion) would certainly be a superior option, if it could be applied – especially if this could be combined with voluntary action by the private sector, aimed at protecting their access of this product to international markets. To date the conventional wisdom on the prospects for this approach has been fairly pessimistic: Oil palm plants do not thrive in the phosphorous and organic matter deficient soils of the large areas of land originally (and to a significant extent still) within the official state forestry boundary – but long devoid of forest cover – and the prospects for significant yield increases beyond those already achieved are widely believed to be weak.

However, there is some recent information which suggests that this pessimism may not be warranted. Large areas of the so-called "vacant" deforested lands within or bordering Indonesia's forest boundary are *imperata* grasslands, called alang–alang, and relatively low-cost operations to add organic matter and phosphorous to these sites has been found to allow oil palm growth at the same rate as on plantations established on newly deforested lands.

Figures in Table 7.1 below show some comparative establishment and maintenance costs and returns for oil palm on alang–alang grassland, secondary forest and sandy soil areas.

Thomas Fairhurst[7], an agronomist with many years of experience with growing oil palm in Indonesia, has found that yield increases on existing oil palm plantations from between 2 and 7 ton of fruit can be obtained just through the application of improved plantation management techniques, with no further requirement

---

[7] Fairhurst, Thomas, pers comm. February 2009.

Table 7.1 Oil palm establishment and operational costs and returns by land type (Fairhurst and McLaughlin 2009. Used with permission)

Indicators	Item	Units	Grassland	Secondary forest (flat)	Secondary forest (sloped)	Sandy soils
Soil type			(Illegible)			Podsol
Planting costs	Planting	$US/ha	3,680	3,950	4,980	4,980
	Fertilizer	$US/ha	1,250	1,250	1,250	1,620
	Increment	$US/ha		+350	+1,150	+970
Operations performance	Maintenance	$US/ha	670	670	755	1,000
	Fertilizer	kg/palm	8.1	8.1	9.1	11.3
		US$/ha	590	590	680	920
	Productivity	t/ha	26.8	26.8	25.5	20.8
Financial performance & planting cycle	NPV(15%)		263	−123	−1,821	−3,888
	IRR		16	14	8	−1
	Break-even	$US/CPO	482	509	667	862

Assumptions: US $1 = Rp 10,000; CPO =US $500.

Used with permission from a presentation to a meeting of the Roundtable on Sustainable Palm Oil, Bali, Indonesia November 2007 by Thomas Fairhurst, David McLaughlin, Haryono, Amri Yahya.

for capital investment. An increase in production of 2 ton per hectare would allow current Indonesian production to be obtained from a plantation area 2 million hectare below the current estate.

A further consideration to be borne in mind here is that if oil palm plantations are located on alang–alang land – assuming that land has been deforested for a considerable period of time as required under present CDM rules – there would almost certainly be a potentially marketable net terrestrial carbon gain from the oil palm enterprise in that case. It is difficult to obtain precise up-to-date estimates of the extent of these grasslands, but estimates from the 1990s put the figure for Indonesia in excess of 20 million hectare.

If the figures given in Table 7.1 on establishment and maintenance costs and palm oil yields in Indonesia are reasonable estimates, then the perceived need of oil palm companies to plant only on recently forested land for purely technical and financial reasons would be called into question.

But which came first – the oil palm or the deforestation? If this is so, then why has so much oil palm plantation ended up on recently cleared forest land?

Deforestation in Indonesia has happened for many reasons: overly heavy (but technically legal) logging, with the forest being subject also to illegal logging before and after licensed logging programmes; softwood pulpwood plantations established under policies which have allowed clear felling of areas considerably larger than has eventually been established under plantation; and other causes. Often, these events have triggered the re-classification of forests under the Indonesian system, from Production Forest or even Conservation Forest categories (which legally cannot be cleared or even, in the case of Conservation Forest, logged at all) to what is known as Conversion Forest – which officially is forest regarded as too degraded to recover as tropical rainforest, although from observations in the field it is clear that much of the forest classified this way is not in fact in that degraded condition.

Even in cases where it is degraded, the question needs to be raised as to whether this was due to an unavoidable series of events on that area of forest, or whether this sequence was in fact allowed, or even encouraged, to proceed by the authorities[8]. Given the previous history of these areas, it would not always be a simple exercise to determine retrospectively what the state of this forest actually was (and therefore what value the logs harvested may have had) at the point of handover to oil palm companies. Nevertheless, Sheil et al. (2009 op cit) are in no doubt on this matter: They argue that historically, areas of forested land allocated under licence for oil palm establishment have been several times larger than the area actually established under oil palm plantation. They attribute some part of this to continued heavy migration to these areas (stimulated to some extent by the existence of the oil palm industry), and to the failure of some areas of plantation. However, they suggest that a large part of it was due to what amounts to timber theft, whereby investors clear the forest ostensibly for the purposes of plantation establishment, but then abandon (or radically scale down) the plantation programme.

This situation has some similarities with the pulpwood plantation situation discussed in Chapter 5: whatever the intent of policy with regard to oil palm development, it seems to have resulted in larger areas of natural forest being removed than could have been planted to oil palm. Since the demise of the Suharto New Order Government in 1997, political forces which could prolong this situation have been in play: decentralization of powers over forests and other natural resources was part of the democratization process that followed the fall of that government: Uncertainties and ambiguities have arisen as to which level of government – national, provincial, or district – has responsibility and authority. Revenue sharing practices established during this period have placed considerable pressure on lower levels of government – especially the District (Kabupaten) level – to generate revenues from exploitation of resources. Given the opportunity, these governments have a strong incentive to liquidate the natural resource capital contained in forests, and convert the land to other purposes. To do so, under present rules, the land must be classified as degraded or Conversion forest, at which point it can be licensed to the Kabupaten for further development – including sale for oil palm development. In short, Kabupaten governments perceive land as having value only after forest cover has been removed.

---

[8] In Indonesia in recent years, adherence to rules of use for land in various forest categories has improved, and some provisions have been tightened. However, there have also been significant changes in the boundaries of the various forest categories, with allocations of forest land to the Conversion Forest category (and thence to licensing for other non-forest uses by Kabupaten governments) in particular growing considerably, at the expense of the Production Forest and Limited Production Forest Categories.

7.2 Stored Forest Carbon: Leading the New Sustainability Paradigm

The bottom line to all this appears to be that, regardless of who has benefited from the process of heavily logging an area of rainforest to the point where it is degraded, the fact remains that if the original site was heavily stocked, then the standing value of the commercially saleable trees on it would have been considerable, and would *alone* present a strong incentive for immediate exploitation. Such a stand might contain 70–80 cubic metres of commercially saleable timber, which with a standing log value of US $30 per cubic metre, would amount to something over $2,000 per hectare. This, combined with the present value of an oil palm plantation then superimposed on the site has appeared to many in Indonesia to be a highly profitable land use combination, in strictly financial terms, and would explain the sequence of events that has been played out so often in Indonesia's forests in recent decades.

Recent developments strongly recommend a re-think on this belief: Firstly, we need to bear in mind that that tropical rainforest can be logged sustainably, and still potentially provide carbon trade value since, if done correctly, this form of logging does not destroy the forest and retains most of the carbon sequestered in its biomass, and in the soil beneath. Some carbon value is lost of course, as some trees are removed (and others damaged in the process). However, Putz et al. (2008) have shown using research data from Malaysian forests that under reduced impact logging techniques that are currently available, and no more costly to apply than conventional logging, an area of tropical rainforest would eventually lose around 60 t of carbon from its original stocking of 200 t or so: the carbon content would recover towards the original figure, but this would take a considerable time to complete. Logging the forest sustainably, which would involve imposition of minimum size limits on trees that could be removed (a requirement which is already included in the forest laws and regulations of most countries with significant resources), might reduce the actual log yield from the forest from 70 to 80 cum per ha to perhaps half this amount; say 40 cum per ha. The net gain from logging the forest to destruction, in log value terms, would therefore be around US $1,000 per ha, over the low impact sustainable logging approach: based simply on halving the volume of logs obtained under a sustainable operation, when compared to an unsustainable logging.

This is likely in fact to be an over-estimation of the difference: In the context of forest exploitation as it has played out in Indonesia in the past, historical standing stumpage values for rainforest logs will underestimate their real market value; poor supervision of legal logging operations, implicit and explicit subsidies on the value of logs for operators, the common presence of illegal logging (which, by definition, pays no stumpage on logs removed) will all combine to drive down the prices received for logs. Under a properly managed sustainable operation, the lower volume of logs being put on the legitimate market will be offset by higher prices for logs (through a combination of the reduced supply, more transparent marketing procedures, and reduced illegal operations).

The second point of relevance here is that as an international interest in forest-stored carbon grows, the economics of preventing the destruction and clear-felling of this forest (which should qualify as reduced deforestation under whatever international provisions emerge under a global carbon trading programme) begin to look very much better: The 140 t of carbon retained on the site would be worth US $2,800, at a (relatively low) carbon price of US $20 per t, or about US $5.45 per t of CO_2.[9] This would be a highly attractive and sustainable alternative to deforestation; and this would be so even if we assume an ample allocation of compensation to re-locating oil palm operations, to offset any additional costs involved in transferring their operations to unforested sites – including the costs involved in coming to equitable agreements with local communities who might presently be using that land for other activities.

The carbon price is, of course, a crucial element in such calculations. Butler et al. (2009) have evaluated returns from various land use scenarios, under differing price scenarios, and in this process concluded that when carbon credits from avoided deforestation are priced at levels obtainable in voluntary markets – which are presently only 10–20% of prices obtainable in a compliance market, such as the European emissions trading market, then avoided deforestation cannot compete with oil palm.

The results from this analysis can be summarized as follows: Development of the concession for oil palm agriculture will generate a net present value ranging from $3,835 to $9,630 per hectare over a 30-year period. Under voluntary markets for REDD operating profits of $614–994 per hectare in NPV over the 30-year period would result- significantly less than profits from oil palm conversion. However, if REDD credits reach price parity with carbon credits in compliance markets, then profitability of avoided deforestation would reach $1,571–6,605 per hectare. This could expand to as much as $11,784 per hectare if carbon payments are front-weighted (i.e. if credits are allocated and sold during the first 8 years when deforestation actually occurs, instead of being distributed over the full 30 years).

The calculations we have carried out above are focused on the prosaic, existing (or, in the case of forest carbon, imminent) markets for goods and services from rainforests. They ignore the other benefits of the forest, from the global and national scale ecological and biodiversity values, through to livelihood opportunities that it provides to local and indigenous communities. As we have argued earlier in this chapter, many of these other goods and services are likely to find more direct market values as time goes on.

The implications of the new information cited above for the potential technical and economic viability of oil palm plantations on vacant lands are profound. It would suggest that the direct opportunity costs for locating oil palm on these lands, rather than on forest, may be quite low, to the extent that the raising of significant sums of money to compensate oil palm producers for locating their operations on the vacant lands may not be necessary. The corollary to this is that retention of

---

[9] What carbon prices may do on world markets in the immediate term is anyone's guess, as the economic crisis continues to influence all markets, and some governments put green investments on hold. However, even in the midst of this, prices (which reached figures of more than € 40 per tonne of CO_2 equivalent in 2008) seem to be holding at around 8 € per tonne of CO_2 equivalent: this is about double the carbon dioxide price used above.

forested land therefore not converted to oil palm would be a highly profitable option for Indonesia, if carbon values from doing so can be realized. Less quantifiable but nevertheless vitally important global and local ecological and social benefits from retaining the forests will also be generated.

7.3 Would Rainforest Governments Finance Sustainability in Forests for Carbon?

We have already noted the essence of what has changed, so far as the fate of the rainforests is concerned given the advent of the climate issue, is the possibility of financing the retention of forests effectively, and the prospect of drawing a wider group of constituents – locally, nationally and internationally – into this effort.

We have also noted more than once in this book, deforestation is a complex phenomenon. While there is general agreement that it is strongly influenced by economic change arising from outside the forest sector itself, its specific causes (and, equally important, its economic and social effects) vary widely between – and even within – countries. In practical, analytical terms this suggests that in some cases aggregate deforestation figures will conflate undesirable forest loss with economically legitimate loss and environmentally benign conversion to other forms of land use, and in such cases the observed aggregate result will have very little policy value. Moreover, deforestation figures do not measure forest degradation unless it progresses to the point of forest loss, but much of the value of a forest can be lost well before that point is reached. Forest production figures are not in themselves an accurate indicator of the condition of forests and forest-dependent people: an increase in forest production may indicate overexploitation of forests, or it may simply indicate an economically desirable and environmentally acceptable outcome such as rising production toward an optimal and sustainable level of output.

The bottom line to this is that we must be wary of attempting to apply economic arguments to make the case for what may in many cases be emotional and highly subjective reasons for protecting rainforests. By all means let us apply economics to the legitimate question of whether – under new market and trading regimes which may be in the offing for certain forest goods and services – sustaining forests and delivering the goods to markets willing to pay for them really is the best approach, from the viewpoint of the owners of those forests (however we define these). However, if we choose to argue, for example, that there is strong will in developed countries to finance the protection of tropical rainforests, then the burden of proof is upon us to show that this really can be translated into financing at the scale required, because it is something that has not happened in the past, despite any amount of rhetoric suggesting that it should happen, and at some (always undefined) moment, would happen.

If we choose to argue that it is in the best economic interests of the country with rainforests itself to protect its forest environment, then we need to be aware of the time frame issue: it is not realistic to expect any government in a poor country in this situation to apply long term low social discount rates to analysis of the case for forest protection, on the basis of qualitative arguments about the value of the forest environment. This is another situation where we need to look for ideas on how to transform the value of sustainable forestry into real and immediate – or at least imminent – economic benefits, not ideological arguments as to the intrinsic value of the forests. If retained natural forests really do now have significantly raised potential economic value, in the form of carbon stored in their biomass, might not rainforest governments want to raise the stakes for those forests, by directing broad economic reform at least partially towards ensuring that those forests do remain intact?

7.3.1 Forest Loss and Broad Economic Change[10]

We have observed in Chapter 2 of this book that many in the international forests constituency have taken a forest-centric view of the world, as if the great stars of economic growth, international and national politics, and social change were orbiting around what, from a vantage point on those large bodies, might look like the very small world of forests issues. However much we in the international forests constituency might believe that forests and related landscapes are important globally – more so now than ever – and that natural resource sustainability is key to the ability of the environment and economies everywhere to maintain our ability to live sustainably on this planet, we must also recognize that a significant segment of the various groups, agencies and others who have major influence over what happens to forests simply do not share this perspective.

Consider the case where a given tropical rainforest country wished to explore the possibilities of using broad economic reforms to build transformation of forest value – in effect, to finance it directly or indirectly through its larger economic programmes – perhaps under the impetus of growing interest in the forest carbon market, along with the wider set of economic goals and objectives that might be involved in the reforms as a whole. One task it would need to undertake, in such a case, would be to examine the relationship(s) between broad economic changes that are proposed, and forest cover.

The first thing that would probably be discovered is that, from the record, defining cause-and-effect in this area becomes particularly difficult when the interest is in estimating the impacts of all large scale economic shifts on forests; obviously, large scale shifts cannot be brought on as part of a controlled experiment to determine what happens to forests in each case, and we need to seek out opportunities to examine the relationship wherever we can. While we can surmise that large economic

---

[10] Much of the material used in this section is drawn from a World Bank report on development policy lending by that organization, and its relationship to forests: one of the authors of this book (Douglas) was closely involved in the writing of this report. The views expressed here are not necessarily those of the Bank, nor of other contributors to the Bank report.

7.3 Would Rainforest Governments Finance Sustainability in Forests for Carbon?

changes in any country, whether through specific reform programs, or through exogenous forces beyond the control of that country, can significantly change the condition of natural resources and the environment, there is no reliable predictive model of what will happen in a given forest situation, from a given economic shift.

One source of information on this can be derived from analyses of large scale and broadly based loans that multilateral development banks have provided to developing countries, in the interests of economic reform and efficiency, currency stabilization, or, more recently, direct objectives in the area of poverty alleviation and in some cases the environment (usually among a much wider group of considerations).

What the World Bank terms development policy lending[11] to developing countries is a classification of loan which addresses issues of poverty and economic reform across a wide range of issues and activities in an economy. Usually this form of loan is preceded by an intensive cooperative assessment of the options and priorities in a given country for poverty alleviation, by the government of the country concerned, the Bank, and involved groups from industry, civil society and others, and the adoption by the government of a reform programme aimed at bringing the necessary changes to policies and their implementation to bear on the goals adopted. In this sense, the resulting loans are quite different to more focused investment lending that the Bank also provides, usually in pursuit of specific project outcomes in specific sectors. They also tend to be much larger than investment loans; usually in the hundreds of millions of dollars, and exceeding one billion in a number of cases.

Development policy loans (DPL) have evolved from what was termed structural adjustment lending by the Bank. These were also large policy-based loans, aimed at economic reform and related issues in a given country. Structural adjustment loans (SALs) have received a great deal of attention from economists, environmentalists, social NGOs and others. They were seen by many as being overly focused on economic development and growth, at the expense of the environment, and of poor and powerless people. Although DPLs seek to address these perceived shortcomings, they are being closely monitored by groups outside the Bank who are concerned to see that they actually do lead to improved poverty alleviation and environmental outcomes.

These large loans (both SALs and DPLs) mimic the sorts of economic changes that might occur in a developing country, as a result of decisions a government might make itself, or of exogenous changes that have macro impacts on that economy. In this sense they provide some insight into what can happen when large and broadly based changes in financial flows occur in these economies. SALs, and now their successors, DPLs, have been more closely scrutinized than economic reforms and changes that have arisen autonomously in developing countries, or than other major economic events which have impacted on those countries.

One of the analytical problems involved here problem is the *temporal factor*. It is well known that the impact of large economic and other changes can resonate

---

[11] World Bank, 2005 *Development Policy Lending and Forest Outcomes,* Report No. 33537-GLB.

through an economy for a considerable number of years, and resulting changes in the natural environment may only become evident at the field level long after the investments made under the reform programmes themselves have been spent. This raises a specific case of the *irreversibility* factor for forests which we raised in Chapter 3: Poor outcomes for agriculture, economic development, or social programmes from policy changes can be identified through monitoring and then in most cases corrected within a reasonable time frame. Impacts causing loss of forests, woodlands and watersheds that depend on this form of vegetation, however, cannot be repaired so easily.

Critics of SALs have argued that the values poor people who live in or near large forest areas derive from those forests are rarely fully incorporated into official economic statistics and decision making. That, of course, would only be a problem if those values are affected by what is done under an economic reform. However, as we will see later in this section, the reality seems to be that the links between broad economic reforms, institutional changes, and policy developments and outcomes at the forest level are likely to be diffuse, indirect, and highly variable from one situation to the next. Therefore, we need to consider some questions:

- What is the evidence that large economy wide changes actually do have significant impacts on forests?
- If the answer to this is that in some cases these impacts are significant, is there a way to predict such impacts, once the nature of the intended change is known?

There are valid conceptual reasons to expect cross sectoral impacts on forests arising from large scale economic change, and therefore to expect that many of the changes in the incentive environment that has frequently accompanied stabilization and economic structural adjustment in the past could have had potent effects on natural resources. Examples of measures that could have had such impacts are: currency depreciation, tighter monetary controls and higher real interest rates, altered trade dynamics and tariff or non-tariff barriers, programs that encourage investment in extensification of agriculture and tree crops without accompanying land zoning and tenure provisions, public sector expenditure reform, and decentralization.

- Currency depreciation can lead to expansion in export of tradable goods in agriculture, tree crops, grazing and expansion in the commercial logging sector itself – all of which can place increased pressure on forest land and resources, although as we have seen in the Indonesia oil palm case study above, perverse sectoral policies and counterproductive incentives can be the real culprits here. The same effect, obviously, can result from increases in the relative prices of the same tradable outputs for reasons other than a currency depreciation.
- High real interest rates shorten optimal forest rotation periods, and tend to increase the relative attractiveness of holding wealth in the form of financial assets instead of natural assets.
- Decentralization policies, which are generally seen as advantageous to governance in terms of increasing accountability and transparency, and are also often introduced on grounds of greater economic efficiency and accountability, could

trigger the irreversibility risk in forests rapidly under conditions of poor sector governance-especially if insufficient attention is paid to the revenue incentives for increasing the rate of forest exploitation that can arise when control over resource decisions is passed to a level of government where other revenue opportunities are limited, and where sustainable and multiple-use forest management expertise is also limited.
- This effect could be compounded by nationally determined public expenditure goals that can further constrain the availability of such expertise. It could be argued that this has actually occurred in recent years in Indonesia, following the introduction of a broadly based decentralization program several years ago.

The World Bank study on development policy lending (World Bank 2005) and forests cites a number of earlier studies of this subject which have attempted to either qualitatively or quantitatively analyse the impacts of International MonetaryFund (IMF) stabilization operations and Bank adjustment lending on forests (see Reed 1992; Repetto and Cruz 1992; Young and Bishop 1995; Glover 1995; WWF 1994; Warford et al. 1994; Munasinghe and Cruz 1994; Persson and Munasinghe 1995).

These studies set out to test the hypothesis that stabilization and structural adjustment programs are harmful to the environment, but the results derived are highly variable. Reviews of them by Dixon (1995) and Panayotou and Hupe (1996) suggest that their results often depend on geographic or sectoral coverage, differences in motivating assumptions, and depth of analysis. Data scarcity and the absence of previous research often forced the authors to make untested assumptions about the causality of adjustment in terms of forest and environmental impacts.

The Bank development policy and forests paper also refers to studies which have focused on a programme of IMF intervention and follow-up Bank adjustment lending implemented in Indonesia in 1998–1999, following the financial collapse in that country (see Seymour and Dubash 2000; Barr 1999; Mainhardt 2001). This remains the only example to date of where both the initial IMF programme and the supporting Bank adjustment operations have included specific forest sector measures in their reform agendas, but again, the findings vary, and definitive conclusions remain difficult to formulate. This is borne out in a review of studies by Gueorguieva and Bolt (2003), which show that the various relationships between the environment and structural adjustment are indirect and complex. Nonetheless, the authors emphasize that there is potential for maximizing positive outcomes and mitigating the negative impacts of adjustment operations on the environment.

Pandey and Wheeler (2001) have used a 38-year socioeconomic database for 112 developing countries, in an analysis of the impacts of structural adjustment on domestic deforestation, and conclude that across the range of situations they have incorporated in their analysis, it is neutral. They note, however, that there is a displacement of domestic deforestation to other countries, and they suggest this is likely in some cases to be a policy concern. Further, their analysis of macro-policy variables reveals that changes in a country's terms of trade does have a significant effect on forest resource use.

Benhin and Barbier (2000) have applied a dynamic optimal control approach to address the forest biodiversity loss issue more directly. They develop a species-forest relationship to explain the link between policy and price changes and forest and biodiversity loss in Ghana from 1965 to 1995 – a period that included adjustment lending activity. In terms of biodiversity, while losses continued during that period, the rate of loss was higher in the pre-adjustment period than the post-adjustment period. The authors conclude that structural adjustment has in this case led to less reliance on forests for production, therefore reducing biodiversity loss.

Shandra et al. (2008) carried out a cross-national analysis of the determinants of deforestation from 1990 to 2005 for 62 poor nations. Their results indicate that the level of national debt, and the presence of structural adjustment programmes correlate positively and significantly with increased deforestation across these countries.

One of the very few studies of the impact of large scale economic changes on deforestation has been done by Wunder (2003). The study examines how changes in exchange rates, government budgets, and consumer spending resulting from oil and mineral exports booms have influenced deforestation. The results reveal that the impact depends on how governments spend the additional revenue, consumer spending, and changes in exchange rate. In Gabon, for example, oil revenue resulted in appreciation of the real exchange rate and growth in non-traded sectors. In contrast, in Ecuador, deforestation accelerated during the oil boom. This is associated with government expenditure of a large share of oil revenues in ways that promoted extensive land use. Also, demand for cattle-derived proteins was important. The study concludes that increases in the export of oil in resource-rich countries will not have a negative impact on forests if labour and other resources are drawn away from the forest and agriculture sectors into the exporting sectors, reducing pressure on forests (see also Wunder and Sunderlin 2004).

In its paper on development policy lending and forests, the World Bank generated a number of internally focused findings from its reading of the literature on this subject and some analytical cross sectional analysis of the matter in its report, that relate to its own future design of DPLs, and due diligence issues involved in their implementation. However, the work also generated some more general conclusions that are relevant here:

- It is clear, from an overview of the studies cited in the paper, that relatively large scale economic changes *can* have significant impacts upon forests. Because these impacts are measured (for the most part) through the highly limited variables of deforestation, or forest output, it is not possible to assert that all such impacts would be classified as bad, from the overall economic growth and sustainability viewpoint, nor indeed from the biodiversity viewpoint – these are matters which are highly context-specific: forest outcomes of major economic reform, targeted at poverty alleviation and sustainable growth, will be highly dependent on the conditions that exist in the sector.
- The high variability of outcomes in forests from specific economic changes suggests that with current data limitations there is little prospect of development of

such a generic model that would allow impacts to be predicted in a specific country situation.
- It is clear that forests are extremely valuable to the livelihoods of large numbers of poor people. It is equally clear that much of this value is not factored into official economic statistics on livelihoods, nor, in many cases, even perceived as value at all. In these circumstances, the risk of inadvertent negative impacts on the poor through economic reform programmes which impact adversely on forests is high.

What we can conclude from this review is that in general terms, it will not be easy to use broad based economic reform – whether internally generated or via development policy lending – as a primary means of implementing a revaluation of existing rainforests: it is simply too blunt an instrument. While such large scale changes clearly can have major impacts on forests, and this should be borne in mind when designing such reforms, other means will need to be found to carry through a revaluation of the forest.

In any event, our focus in this context is the policy implications of the observed results. A positive correlation between the presence of an economic adjustment operation and the level of deforestation does not indicate that that adjustment activity should simply be removed: what it suggests is that where the overall aims of the adjustment are being achieved, and have high priority, the answer to a particular negative effect also being generated – such as increased deforestation – is to offset that effect: In situations where the value of the forests affected is seriously underestimated – which as we have suggested is often the case – then measures to insulate the forests from the broad deforesting dynamic that economic reform has initiated is a priority. In our view, these will involve ways of directly investing in protection of the forests, including financing of the necessary incentives to ensure that all potential agents of deforestation are dissuaded from such activity. And this, of course, will require approaches that can access the large amounts of funding that will be needed (see the discussion earlier in this chapter). We suggest that this in turn will involve approaches that can access the elements of the international private sector with a market interest in the forest carbon issue, in addition to whatever resources can be drawn from the public sector in developed countries.

7.4 Financing Reduced Emissions from Deforestation and Forest Degradation

In our consideration of the options for financing forest sustainability at an effective level, we have now eliminated two potential sources of such financing:

- We have argued in Chapter 6 that direct developed country donor financing of the full costs of significantly reducing deforestation, has historically been inadequate for the task, and is likely to remain so.

- We have argued immediately above that attempting to offset pressures on forests via broad economic reform programmes is likely to be difficult to focus appropriately. It is also likely to carry significant political risk to governments which are heavily dependent for revenues on agricultural enterprises and forest based industries – both sources of deforestation pressure under economic expansion regimes.

This brings us back to the subject of the market for carbon sequestered in tropical forests. Forest based sequestered carbon is unique among what can be termed the forest public goods, in that while retention of carbon in forest biomass has the public good characteristic of being of benefit to everyone, regardless of who covers the cost of providing this benefit, it is also a product which can be sold to the private sector without compromising its public good benefit. A country which can provide sequestered carbon via reduced deforestation will in all probability fairly soon be able to obtain payment from international sources for doing so: this is the driving force behind the reduced emissions from deforestation and forest degradation (REDD) initiative.

On the other hand, for biodiversity and the other public goods, there is little prospect in the immediate future of a large international scale commercial market developing for these products. If they were to be considered as separate from the forest carbon product, they would need to rely upon essentially grant-based international financing for some time, for there to be any chance of their being made available from rainforests perpetually. As we will see, there are possibilities for having them regarded as joint products along with the value of carbon retained in forest biomass. Governments (or possibly in some cases large private sector entities) which provide this financing may choose to regard this as their purchase of global forest public goods, rather than conventional donor financing of development in a given country, but the point remains that there is unlikely to be a commercial market in which private agents trade in these goods for their own financial benefit, as will probably soon be the case for forest carbon, and therefore the REDD option is going to have important implications for the other important global public goods which rainforests provide.

7.4.1 REDD Has Been a Long Time Coming

As many readers will know, a formal programme for design and implementation of the REDD initiative under the auspices of the United Nations Convention on Climate Change (UNFCCC) took quite a while to get under way. The Kyoto Protocol was adopted by the parties in 1997, but did not come into force until 2005, when sufficient nations were signed up to the commitment.

REDD, along with other activities and agreements under the UNFCC, will recognize the differing status and responsibilities of various national partners on GHG reduction. The Convention divides countries into groups on this basis, as shown in Box 7.1 below.

> **Box 7.1 Recognized parties under the UNFCCC**
>
> **Annex I** Parties include the industrialized countries that were members of the OECD (Organisation for Economic Co-operation and Development) in 1992, plus countries with economies in transition (the EIT Parties), including the Russian Federation, the Baltic States, and several Central and Eastern European States.
>
> **Annex II** Parties consist of the OECD members of Annex I, but not the EIT Parties. They are required to provide financial resources to enable developing countries to undertake emissions reduction activities under the Convention and to help them adapt to adverse effects of climate change. In addition, they have to "take all practicable steps" to promote the development and transfer of environmentally friendly technologies to EIT Parties and developing countries. Funding provided by Annex II Parties is channelled mostly through the Convention's financial mechanism.
>
> **Non-Annex I** Parties are mostly developing countries. Certain groups of developing countries are recognized by the Convention as being especially vulnerable to the adverse impacts of climate change, including countries with low-lying coastal areas and those prone to desertification and drought. Others (such as countries that rely heavily on income from fossil fuel production and commerce) feel more vulnerable to the potential economic impacts of climate change response measures. The Convention emphasizes activities that promise to answer the special needs and concerns of these vulnerable countries, such as investment, insurance and technology transfer.
>
> The 49 Parties classified as **least developed countries** (LDCs) by the United Nations are given special consideration under the Convention on account of their limited capacity to respond to climate change and adapt to its adverse effects. Parties are urged to take full account of the special situation of LDCs when considering funding and technology-transfer activities.

Even by the time the Kyoto Protocol was finally implemented there had been no agreement under the protocol to allow for carbon emissions trading by means of reducing deforestation in developing countries. Carbon credits were allowed for Annex I countries implementing projects in partner Non-Annex I countries, to sequester carbon from the atmosphere by specific land use changes, such as reforesting or afforesting areas of land, under the provisions of the Clean Development Mechanism[12].

---

[12] Apart from being restricted essentially to forestry options based on reforestation and afforestation, rather than avoided natural forest reforestation, the CDM has been criticized for a complicated verification and implementation framework, and it has not been widely utilized ion many developing countries since its inception.

The restriction on recognition of avoided deforestation as a legitimate carbon offset trade mechanism was due in part to the fact that at the original Kyoto discussions, little preparation had been done on the difficult issue of benchmarking in this situation, which would be needed to be certain that real gains in carbon sequestration or prevention from emission in the case of already stored carbon were being made by a given activity. Oberthur and Ott (1999) have described the situation which prevailed at the time, and have commented that this issue was of the greatest importance, with serious implications for the credibility, transparency and verifiability of the emission targets and with great impact on the eventual size of those commitments.

It was also the case that a number of country governments at the time, along with some non-governmental environmental organizations were concerned about the idea of reduced deforestation as a carbon offset trade. Some (for example, the Government of Brazil) were concerned that placing large areas of rainforest under a trading regime of this nature might threaten national sovereignty over this resource. Others had doubts about the potential moral hazard in using a current high rate of natural forest loss in a given country as a benchmark against which future reduction would be measured, and argued instead that protection of existing forest estates should be a "normal responsibility" of governments, in order to achieve long term protection of their natural resource bases; others had concerns that a flood of cheap forest carbon credits onto Annex I country markets could be a disincentive for industries and other high emitters in those countries to get on with the task of directly reducing their own emissions. The sovereignty issue appears to have largely disappeared from the debate recently (at least, for the time being), but others among those listed above remain contentious, and we will discuss these later in this chapter.

The majority of participants in the Kyoto process did accept in 2005 that avoided deforestation in Non-Annex I countries should be allowed as an option under Kyoto – and initiation of the REDD process by which this is to be achieved was endorsed by the parties at the Committee of the Parties Meeting number 13 (COP 13) at Bali in 2008. Overall, since the original Kyoto meeting, the dynamic towards support for inclusion of reduced deforestation as a legitimate offset under a trading regime seems to have grown.

A graphic indication of this is given in the approach developing in the United States under the Obama Administration. At the time of writing of this book (mid-2009) the Waxman–Markey Bill which had passed through the Congress and was on its way to the Senate had committed the Administration only to a very modest target of reducing emissions to 17% below 2005 levels by 2020: In effect, this would reduce US emissions by that date to about 1990 levels, whereas the European Annex I countries were talking about targets for 2020 of 20–30% *below* their 1990 emission levels. However, President Obama had raised the prospect of expanding the US targets significantly, by means of global investment in reduced deforestation in Non-Annex I countries. The Australian Government has made similar statements about the prospects of investing in reducing deforestation in these countries as a means of offsetting high emissions from domestic Australian industries and power utilities in that country.

7.4.2 The Basics of REDD

A section from an excellent guide book on REDD prepared by the Global Canopy Programme is excerpted in Box 7.2 below.

Note that Global Canopy is careful to avoid linking REDD directly to a market system for trading forest carbon offsets in Annex I countries, and as we will see later in this chapter it is appropriate at this point to retain this flexibility in approach. Some advocates (and indeed opponents) of REDD do appear to conceive of REDD as primarily a system for placing forest carbon offset trades from avoided deforestation into some sort of formal market, where forces determined by the demand among high emitters for offset options to augment whatever they may be required to do the reduce their own emissions directly will determine the price paid to Non-Annex I forest nations. This is definitely an option under consideration, and it may well be that eventually it develops into the most intensively applied

---

Box 7.2 Outlining REDD

What is REDD?
The basic idea behind Reducing Emissions from Deforestation and Degradation (REDD) is simple: Countries that are willing and able to reduce emissions from deforestation should be financially compensated for doing so6. Previous approaches to curb global deforestation have so far been unsuccessful, however, and REDD provides a new framework to allow deforesting countries to break this historical trend.

What are the objectives of REDD?
REDD is primarily about *emissions reductions*. The Bali Action Plan decided at the Conference of the Parties (COP) at its thirteenth session 7 states that a comprehensive approach to mitigate climate change should include: *"Policy approaches and positive incentives on issues relating to reducing emissions from deforestation and forest degradation in developing countries; and the role of conservation, sustainable management of forests and enhancement of forest carbon stocks in developing countries."*

More recently, the "+" in REDD+ has drawn increasing attention towards the activities after the semicolon, related to the conservation and enhancement of carbon stocks. A future REDD mechanism has the potential to deliver much more. REDD could simultaneously address climate change and rural poverty, while conserving biodiversity and sustaining vita ecosystem services.

Although these benefits are real and important considerations, the crucial question is to what extent the inclusion of development and conservation objectives will either promote the overall success of a future REDD framework or complicate and therefore possibly hamper the ongoing process of REDD negotiations.

(continued)

> **Box 7.2** continued
>
> **The story so far...**
> A fundamental milestone was achieved at COP 11 in Montreal in 2005 when Papua New Guinea and Costa Rica supported by eight other Parties proposed a mechanism for Reducing Emissions from Deforestation in Developing Countries. The proposal received wide support from Parties and the COP established a contact group and thereafter began a 2 year process to explore options for REDD. This decision resulted in a wide range of Parties and observers over this period submitting proposals and recommendations to the Subsidiary Body on Scientific and Technical Advice (SBSTA) to reduce greenhouse gas (GHG) emissions from deforestation and degradation. We are now at the stage where we have a number of proposals on the table. Under the Bali Action Plan, if REDD is to be included in a post-2012 framework, a decision about what a REDD mechanism will look like and what it will include needs to be agreed by COP15 in Copenhagen in December, 2009. Reaching a consensus on this issue is of paramount importance for a global deal on climate change.
>
> *Source*: Parker et al. 2008

approach. But it is important to bear in mind that the basic principle of REDD is the delivery of financial compensation from Annex 1 countries to provide incentive for Non-Annex I countries with the potential to reduce deforestation significantly to do so. The *mechanism(s)* by which these transfers occur are not set in stone in this point.

Given this, and the fact that significant doubts and disagreements remain amongst some observers of and participants in the REDD process about how it should function, we need to consider some of the issues that are feeding these trepidations.

7.4.3 Issues and Differences for Consideration Under REDD

We will begin by outlining some of the matters that are presently being discussed in the dialogues and literature on this subject, because these will form the basis of much of the debate which will take place at the UNFCCC COP 16 in Copenhagen on the REDD issue, and they will also feature largely in the negotiations of specific arrangements and rules for REDD which will be negotiated from then on. Readers will understand that the global level of discussion on REDD is intense, fast-moving and volatile, and the literature on this subject is enormous, and expanding quickly. Under these circumstances, anything we say or report here runs a high risk of being

7.4 Financing Reduced Emissions from Deforestation and Forest Degradation 177

out of date or irrelevant in a fairly short space of time. We will attempt to confine ourselves to some of the more central issues, which at least will assist in interpreting future developments in this area, to some extent.

Some broadly based doubts At risk of oversimplification, there are two basic categories of argument surrounding the REDD approach that seem to be of primary concern to many of the interest groups watching this debate at the present time. The first questions the focus of REDD; in particular, whether it might create adverse impacts on particular groups in society. The second questions whether REDD as presently conceived by its protagonists can actually work at all.

An example from the first category is provided by The Forests Dialogue group of organizations, which has published a number of articles from various of its members on this subject. The overall thrust of these is captured in a statement by TFD itself (TFD undated) on REDD, which contains the following commentary on current plans for REDD:

- REDD will need to develop an approach to emissions reduction that will effectively involve countries which do not yet have high rates of deforestation, as well as those that do.
- The present focus of REDD on a single commodity – carbon – instead of multiple forest values is unlikely to succeed, and will undermine the social, environmental and economic resilience of rural communities and indigenous groups.

TFD observes that the main drivers of deforestation lie outside the forest sector, and must be dealt with as such. It offers some guiding principles that are aimed at ensuring that whatever is done under REDD gives impetus to sustainable development, improves governance, strengthens land tenure and carbon rights, and finances the building of capacity of countries, communities and forest managers to participate effectively in climate change mitigation and adaptation strategies.

An example of the second major line of argument against mainstream views on REDD – the argument as to whether REDD as presently conceived is feasible at all is provided under the auspices of The World Resources Institute, which has published a draft summary (Daviet et al. 2008) of a longer paper it is intending to issue. This summary poses some issues which go to the heart of the feasibility of REDD, as presently conceived. It suggests that the focus in present thinking in REDD on national level metrics for reduced deforestation may be counterproductive:

- It will exclude from consideration worthwhile project level activities which have achieved forest-saving in advance of being able to monitor and report national level gains or losses in forest area (e.g. declaration of protection status over a valuable piece of forest).
- If the REDD criterion for payment and recognition of carbon benefit is simply a net national reduction in deforestation, this alone will not guarantee the survival of the whole forest estate, and the carbon stocks it contains.

The suggestion in the paper is that a two track option might be preferable – one focused on funding positive conservation projects and programs, and another that would track the rate of deforestation.

The authors suggest that three fundamental changes (or perhaps additions) to the current REDD approach are necessary:

- A dedicated fund for REDD, rather than an offset mechanism linked to GHG emissions. WRI accepts that there is a risk attached to the question of whether Annex 1 governments would finance such a fund sufficiently, but counters by suggesting that the transaction based approach may not raise large transfers either, since at present neither the US nor EU markets have yet shown any appetite for such credits (but note that the recent statement by President Obama on this matter referred to earlier in this chapter suggests this may be changing).
- REDD support for the Sustainable Development Policies and Measures (SDPAMS) approach, which in essence argues that developing countries need to develop themselves before committing to activities aimed directly at climate change: this (presumably) will make them better equipped to deal with the climate change issue eventually.
- Global supply and demand programs, aimed at reducing sales of unsustainably produced outputs from forests (i.e. overlogging) or from land cleared of forests.

The paper finishes with a detailed list of questions and issues which it suggests must be addressed before REDD can be finalized under the UNFCCC.

There are some major issues associated with REDD which we need to outline here. The UNFCCC has itself commissioned major pieces of work, some on specifying what the methodological, financial and institutional questions and issues to be addressed should be, and more on the actual analyses required to answer some of the questions. The key issues the UNFCCC acknowledges will have to be addressed are classified in the CIFOR/IPAM/ODI (2008) background paper under the broad groupings of methodological, financial and institutional subjects:

Methodological issues: The scope of accounting systems that should be applied to REDD could account for deforestation alone, or attempt to include the more difficult question of forest degradation. Degradation is frequently a predictor of future deforestation, but it is also more difficult to measure in the field. There is also the question of whether REDD should be considered as part of a much broader suite of terrestrial carbon sources and sinks (which also raises the question of the impacts of different vegetation on soil carbon, and how (if) that should be credited. Additionally, the matter of whether the basis of evaluation should be maintenance of forests in a steady state, as some ecologists and others argue – but which would imply a great deal more than simply reducing deforestation – will need to be addressed: we discuss this issue later in this chapter.

Baselines against which deforestation can be evaluated are going to be necessary, but are problematical. Baselines calculated on historical rates of deforestation are dependent on the availability of data and its accuracy. If applied widely, or globally, some commentators point out that such baselines may run a risk of acting as a disincentive for countries with low rates of current deforestation from maintaining their forests in this condition, and as a "reward" for past poor performance in high deforestation countries, and the baselining exercise might result in an exaggerated projection of where deforestation is headed. Alternatives will need to be considered, especially in cases where a reasonably reliable historical record of deforestation may simply not exist. Future deforestation can be projected on the basis of modelling, using common factors believed to be causal in deforestation; targets for reduction can be determined through negotiation for specific countries; and it has been suggested that world average deforestation rates could be applied as benchmarks in some cases.

In the case of Annex 2 countries with significant natural forests and low deforestation, the central issue is that history suggests very poor nations with large forest resources will, sooner rather than later in most cases, begin to liquidate the value of those forests by heavy logging and then conversion to high value crops. As discussed in Chapter 5 of this book, the literature surrounding the environmental Kuznets curve theory suggests that it will take very large movements in per capita income before the resulting deforestation begins to slow down in a given country, as different environmental perceptions of forest value begin to develop: Malaysia is usually offered as the example here: deforestation began to slow in that country around 10 years ago, by which time Malaysian per capita income had reached something like ten times the equivalent in neighbouring Indonesia, and the great majority of the natural forest on Peninsular Malaysia was already gone.

The problem is, it is difficult to project when a given rainforest country with a currently low deforestation rate will begin to ramp up this rate. Hence, it is difficult to determine a realistic opportunity cost schedule through time against which a compensatory payment could be made to provide a sufficient incentive to deter the deforestation process. There are some reasonably obvious alternative approaches to valuing the retained forest which could be considered:

- As noted earlier in this chapter, some recent scientific publications show that mature rainforest in fact is absorbing carbon, not emitting it. There are figures suggesting that this could amount to 1 or 2 t per hectare – and bear in mind this will be for every hectare in the forest, every year. Since carbon is now a market commodity, it is reasonable to suggest that, if this research is confirmed, this forest carbon absorption is something the global community could well consider purchasing, in a low deforestation country. A deal could be struck, whereby the rainforest country undertakes to maintain its forest estate at or above some negotiated threshold level, in return for payment for the value of carbon being absorbed by the entire estate. Obviously, the payment figure would need to be sufficient to offset the alternative uses of forested land (plus the value of rapid, non-sustainable utilization of the forest) that are expected to otherwise occur.
- An attempt could be made to directly estimate what the potential opportunity cost to the rainforest country of not beginning to raise deforestation will be.
- A third, more difficult way to approach the problem, would be to determine from macroeconomic figures and precedents what the level of new capital to stimulate an acceptable level of economic growth for the country would be, and to calculate what level of funding would be needed to build existing amounts to this level, on the grounds that this might be something similar to the amount that would have been sought for investment in oil palm or other crops that might encroach on forests.
- One possibility which has been discussed would be to assign a notional rate of deforestation to such countries, and pay an incentive as if that rate were being reduced to zero: this would, in effect, be a levy on the funds to be invested in REDD to produce reductions in high deforestation countries.

One way or another, it should be possible to develop an incentive framework which rewards continued low deforestation. Or, more simply, international donors could allocate sufficient grant based financing to such countries, on the basis of supporting sufficient agricultural and other development programmes to take up the population and other pressure that would otherwise be driving the dynamic towards deforestation. As discussed in Chapter 6, this level of funding has not happened under regular donor programmes to date, but assuming there is some agreed notional valuation of the carbon benefits of doing so now, more intensive efforts might be stimulated.

For both the high-deforestation and low deforestation scenarios, an important methodological consideration is that the calculation of the carbon outcomes of deforestation is not straightforward: there are not yet good data on the amount of forest biomass on all areas where deforestation is occurring. Deforestation itself can be measured at large scale relatively cheaply via remote sensing, but the same cannot yet be said of actually estimating the amount of biomass which is on those deforesting sites. A deal more information on forest carbon stocks around the world is going to be needed in this case.

A much-discussed issue under the general heading of methodology is that of whether national approaches to REDD are to be the standard, or whether sub-national projects in isolation will be recognized. The primary risk associated with sub-national projects is known as leakage: reducing deforestation in one area of a large national forest estate might be equivalent to a net reduction on deforestation for that country as a whole, but it might equally simply be displacing deforestation from one region of the forest resource to another. Nationally coordinated REDD programmes would require a national assessment and would therefore avoid the leakage problem, but they could also reduce participation, especially among countries with limited capacity and financing to implement a programme at national scale. In all probability, the REDD negotiators will need to come up with a compromise on this; one which allows certain countries to implement project scale REDD activities for some period of time (and receive recognition under the REDD rules for these), with the intention that these arrangements will eventually give way to a nationally coordinated programme.

We also note here that the problem of dealing with low-deforestation rainforest countries, as discussed above, can be seen as an extension of the leakage argument to the international level: it is conceivable that where a high deforestation country reduces its levels of deforestation, increased pressure on supply of logs from presently low-deforestation countries could be the result. Further, if a high deforestation country is reducing its plantation of high value cash crops onto forested land, and is incurring opportunity costs of re-locating those planting in doing so, then there may also be a response in low-deforestation countries to begin establishing those same high value crops on their forested lands, unless there is an existing incentive not to do so.

Financial issues. There are two basic approaches to financing a REDD programme in a given country – or indeed globally. The first is market-based approach, where the objective is a trade of REDD credits between a national seller, and a buyer – either an Annex I country, or corporations within Annex I countries. The second is

7.4 Financing Reduced Emissions from Deforestation and Forest Degradation 181

a fund-based approach, where international grant-based funds sourced from donors (Annex I governments, or other large entities capable and willing to provide significant funding) is used to pay Non-Annex I providers of REDD benefits via recognized reduced deforestation are paid according to their efforts in this regard.

The market based approach has raised a number of questions: Some participants on the REDD debates have noted that allowing significant volumes of REDD credits onto some Annex 1 markets could act as a disincentive for high emitters in those countries to invest in reducing their own emissions directly (presumably because they expect REDD offsets to be less costly than direct emissions reduction in the Annex I economy itself). Others believe that the market approach may be accompanied by perceptions on the part of potential investors of significant risk, either that the REDD programmes will not function effectively, or that at some point the supply of REDD trades may simply be cut off by supplier governments or projects. If these sorts of perceptions do manifest strongly in the market, they would create significant up-front costs for potential suppliers of REDD credits, before potential buyers are convinced that the process will work satisfactorily, and meet all standards. There may, as a result of potential issues such as these, be some tendency in some Annex I countries to negotiate for limitations on the supply of REDD credits that might be allowed into that country's market, or for temporary "banking" of credits until the flow is more regularized. A question raised by some, in view all of this, is whether trading for REDD credits should be allowed to take place in mainstream emissions trading markets at all, or whether it might be preferable to establish a parallel market for the REDD trade. We discuss the issue of "market flooding" in the sub-section on controversial issues which follows.

The major problem foreseen for the funds based approach harks back to our discussion of donor funding for sustainability, discussed in Chapter 6 and earlier in the present chapter of this book: the gap between the costs of financing effective reduction of deforestation, and the historical amounts the international donor community has provided for sustainable forest management and forest conservation is very large; even when new donor initiatives passed around capacity building for REDD are factored in, the gap is still large. Some commentators have expressed fears that raising donor commitments to REDD significantly would reduce ODA commitments to other important international programmes. The CIFOR/IPAM/ODI (op cit) background paper also notes that the funds based approach might be less performance-based, and less concerned about risk and efficiency than market-based approaches. Certainly, from our own consideration of the effectiveness of donor and other grant based financing of sustainable forest management and forest conservation in this book, we would be inclined to concur with this assessment.

Institutional issues and capacity building: Credible verification systems will be required to verify estimates of emissions reductions, and changes in cover, and they will also be needed to ensure accountability in the financial mechanisms which are developed to implement the REDD processes in partner countries. How broadly these systems will function, how they will be managed, and how they will be adapted to specific national situations are matters which have not yet been defined. There will need to be some consistency of standards built into the systems across all countries, and there is a possibility that this might re-ignite the sovereignty issue, in some cases, unless appropriately handled during the negotiation process.

Significant capacity building investments will be needed to bring most rainforest countries to a state of "REDD-readiness": Inventories of deforestation; monitoring and reporting capability; establishment of an enabling environment for trading will all need to be effective at an acceptable standard, and in some cases significant reforms in issues of sector governance, tenure and access to forest lands by various interest groups will be required to make this possible. The CIFOR/IPAM/ODA background paper expresses some doubts that either market or fund based approaches to funding the REDD process will provide sufficient funding to allow these developments to take place, therefore increasing the risk that some potential supplier nations may not participate. We suggest that the figures cited earlier in this chapter relating to capacity building costs demonstrate that these are going to be highly significant; if the funds to support an adequate effort cannot be found within budgets for market or fund-based REDD programmes per se, then it will need to come from elsewhere in Annex I country budgets.

We take the points made under the finance heading earlier that there will be a need for significant investment in capacity building before strong commitments to REDD under whichever financing option is chosen will be made. It would be better to regard these as clearly phased and contingent activities, with the needed investments in capacity building being committed and effectively disbursed before any significant purchase – even forward purchase – of REDD trades or incentive payments will take place.

7.4.4 Some Controversies and the "Agenda Loading" Issue

As will already be apparent from the foregoing discussion, there remain many issues and questions which remain un-addressed at the current stage in REDD development; some of these are mainly technical, others are more political. In the latter group, we can see that the historical tendency for some interest groups to attempt to obtain special consideration for a particular issue in international initiatives – which we have remarked upon as a common phenomenon in the international forests constituency – has flowed across into the deliberations on REDD rules or conditions. In some cases (for example, the proposal that Annex I countries might limit the access of REDD trades to their markets, in order not to discourage progress in their own internal emissions reduction programme), the objectives of such proposals are decidedly focused on domestic political issues.

In other cases, the concerns which have been raised are based upon worthy global objectives for forest management. These include the need to assign a high profile to the dependence of the poor on forests for basic livelihood; the necessity to provide actual legal tenure over forested areas to local communities; the need to place biodiversity conservation at the centre of activities carried out under REDD; the desirability of linking REDD financing to campaigns to exclude non-sustainably produced commodities from Annex I markets, as discussed in Chapter 5 of this book, are examples. Worthiness of intent does not, however, guarantee that including specific goals and targets on these sorts of things into REDD agreements will achieve beneficial outcomes, rather than simply adding to the already formidable list of risks and constraints on implementation of REDD.

We do not seek here to argue against the implementation of measures to improve upon outcomes in areas of global concern, nor to belittle their importance. We do suggest, however, that the maintenance of as much forest in an intact and viable state is an overriding objective here: The outcomes for poverty alleviation, community participation and so on being sought by those advocating incorporation of hard and fast conditions into REDD to achieve them will not be achieved if REDD does not succeed in its basic goal. The more rules and conditions a provider of a given REDD project is expected to meet, the less likely it is that that project will succeed, no matter how desirable achieving them might seem. Sometimes, the price of doing no harm is set so high as to guarantee that projects which can satisfy all pre-conditions will also do no good.

Given the difficulties that REDD will impose for its implementers everywhere just to meet basic requirements on reduced deforestation, we need to be especially vigilant that agenda-loading, and allowing the perfect to get in the way of the good, are avoided. We discuss three issues in this regard below, to offer more specific examples of the point we seek to establish here.

Communities, poverty and the tenure issue: There is significant debate around the forest tenure debate as it impacts on REDD. On the one hand, some social and environmental civil society groups argue that without effective tenure for local communities, and indigenous groups, in forests where there is significant competition for the forest resources themselves and for the land beneath those forests, measures to reduce deforestation will not succeed: instead, a perpetuation of long-running conflicts over land and resources, and of inequitable and frequently corrupt practices in the allocation of that land, will prevent any progress. In similar vein, the large numbers of poor people who – whether they may have some historical association with the forested areas they now depend upon for livelihood or not – need to be offered some form of secure tenure to relieve their dire situation.

On the other hand, some policymakers and potential financiers of REDD type activities will argue that establishing actual new tenure arrangements over what is, frequently, land and forest which has for better or worse at some point in the past been effectively appropriated by the state, will of necessity be an extremely long drawn out process. If the REDD development is held up until it is completed, then there will be no hope of REDD fulfilling its promise as a means of re-valuing rainforests. The history of land tenure changes to include the rights and traditional local community ownership of land shows that it is an extremely time-consuming business, likely to take decades to complete. Thus, the tension between those who advocate actual tenure change as a prerequisite for REDD, and those who argue that the very essence of REDD is speed – that the longer it is delayed, the more forest is lost and the less useful it will be – is a genuine and serious conflict.

We are aware that tenure reform in forests is possible: As has been pointed out by The Forests Dialogue group, significant progress has been made in community ownership of forest areas in recent years: this has doubled, to a total area of 377 million hectares – about 22% of all forest area in the developing country group. This trend continues, and some analysts estimate that this figure could reach 50% within the next two or three decades. The literature on community forests cites many examples of the improved incentives for sustainable management that are transferred to communities when they receive rights to utilize forests for their own benefit, rather than having to exploit them illegally, as is the norm when tenure and access issues are less favourable (see White and Martin 2002).

Nevertheless, large problems remain in the area of community tenure in forests. In some countries (especially the Melanesian group of countries in the South Pacific) clear land ownership has been vested in clan or community groups since these countries achieved independence, but the reality is that this has not resulted in a genuinely equitable sharing of the benefits of use by others of these forests, and nor has it resulted in sustainability of these forests. In other countries, problems of land being heavily contested by rival local communities or interest groups can threaten internal stability when the question of assigning tenure arises. In some rainforest nations, the ownership and permissible use of land is disputed even among the levels of government, from national, through regional, state or provincial, down to local levels.

A compromise solution may well be the only solution in such cases, where it is vital that the rights of access and usage of forest areas for local communities, indigenous groups and other poor people be maintained and developed, and it may also become important that some form of carbon rights be assigned to these groups as well, so that consensus and co-operation, based on development of equity, will replace conflict in these cases. In fact, in situations where community-government relations over land access and use issues have been fraught in the past, it is difficult to see how this could be allowed to persist, if the government and the communities themselves really do intend to benefit from international trade in the goods and services that can be produced from intact rainforests. As is well known, sustainability of large natural systems, especially in heavily populated countries, depends on relationships of trust and mutual benefit between governments, communities and private investors.

Biodiversity protection, permanence, and maintaining the existing forest stock: The case has been made by some for targets for REDD to be set on the basis that no further deforestation should occur in a country which wants to receive REDD financing and market opportunities: elements of this approach can be seen in the World Resources Institute paper reviewed earlier in this chapter. The argument is that the idea of simply slowing deforestation does not guarantee that there will be any natural forest left after a period of time, therefore the REDD programme will not in such a case have protected forest biodiversity into perpetuity. Time-limited contracts for REDD are seen to suffer from the same flaw, in that if a given agreement between a REDD trades supplier and its purchasers covers only a few years, then there is nothing to prevent that supplier from resuming rates of deforestation at the previous higher level when the agreement expires.

From a purely economic perspective, the value of carbon retained from reduced deforestation, and sold to the market as such, would be valid even if, after a few years, deforestation rates returned to previous levels: so long as deforestation levels did not move to levels *above* those prevailing before the reduced deforestation regime was introduced, then any carbon reduction obtained during the period of reduced forest loss has made a genuine contribution to reduced global GHG emissions for that period[13]. However, if the objective is to create and protect biodiversity reserves, or to maintain forested catchments, or similar objectives, then a trading arrangement that promised a life cycle of just a few years would not be regarded as a useful exercise.

Ideal as it would be to achieve the larger biodiversity protection goals forever, via REDD, we need to look at the risks such a requirement would pose. Reducing deforestation will be extremely difficult for some rainforest nations, and it will occur by progressive reductions: no such country would be at all likely to enter a REDD agreement that offered no payment for results until the point of zero deforestation is reached.

Could forest carbon trades "flood the market?" As noted earlier, there is an intensive discussion going on in the international forests community, and more broadly in the emissions reduction dialogue, on the possibility that a large amount of "cheap forest carbon trades" might flood the nascent markets of Europe, thus reducing incentives for investment in higher cost emission reduction options there. There has been discussion within REDD circles and elsewhere that a limit might have to be placed on the volume of forest carbon trades that would be allowed onto the European market.

The first point to be made here, in light of the rough calculation of the competitiveness of carbon from forests outlined above, is that it is not certain that forest carbon in large volumes will be able to be put onto global markets competitively at this point, and a lot more analysis of the options for (and costs of) re-directing cash crops from forested lands elsewhere will be needed to establish the case one way or the other.

The second point is to note that the room in the market for forest carbon, as for any other emission reduction option, will be determined by the intensity of the emissions cap that Annex I countries apply to themselves in the post-Kyoto 2012 agreement. The World Wildlife Fund (2008) has provided some analysis of this, and the following indicative figures are extracted from the results:

- A 90% reduction in global deforestation would generate 3.2–6.4 Gt of carbon dioxide reduction. This would be four to eight times the amount of the present annual target of the Kyoto Protocol at present.

---

[13] Unless it can be argued that the country has chosen to revert not merely to previous higher levels, but to even higher levels than would have occurred under a business-as-usual scenario without REDD. That would be an abrogation of that nation's understanding with the REDD partners; it could happen, but then, it could happen at any point under any form of agreement. There is an element of faith in the REDD process: it assumes that nations which participate in the process will see the benefits – economic, social and environmental – from being in the programme.

- If the EU adopts a 30% emissions reduction target, and chooses to achieve one third of this goal through purchasing credits from outside the EU zone, then the total of that external supply of credits would amount to 493.3 megatonnes (Mt) of CO_2: a much smaller figure than the potential emissions reduction from significant deforestation reductions at the global scale.

It should be noted here that these figures seem to compare annual emissions targets for the EU, with *total* carbon reductions from reduced or eliminated deforestation – but these latter figures are essentially once-off amounts. Once achieved, there will be no further credits available. It is much more likely that carbon credits will arrive on markets over an extended time frame (see the 10 year span used in calculations referred to earlier), and therefore the amounts arriving on markets in any year will be much lower than the aggregate numbers cited above.

It would seem, from the above, that carbon trades from reduced deforestation are not likely to flood large global emissions trading markets, and nor are they likely to persist in those markets for an appreciable time: the deforestation calculations we have cited in this chapter are mostly based on an assumption that rates of deforestation in the Non-Annex I countries could be reduced to very low levels in 10 years. Were that to be the case, then beyond 10 years the supply of reduced deforestation would be very much reduced. In view of this, it would be unfortunate if the (apparently intensifying) discussion of limiting access to emissions trading on the European market de-railed the current interest in the deforestation option in potential supplier countries, before it has really even had an opportunity to fully analyze the options and potential that may be there.

7.5 Investing in Reduced Deforestation Ahead of REDD

One thing upon which most observers of the developing REDD process can agree is that for implementation of reduced deforestation, time is of the essence. UNFCCC-REDD sanctioned solutions could well take another decade before actual field programs under REDD conditions are in place, especially if rainforest nations wait until new post-Kyoto provisions are formalized into an international agreement in 2012, before even beginning to invest in REDD readiness: this may be the case with some of the presently lower deforestation group. The task of designing a global market instrument for forest carbon offsets, then setting in place the rules, the capacity to implement them and the infrastructure to measure and monitor the whole process is an immense one.

Recognizing this, the UNFCCC itself has called for early implementation of reduced deforestation options, on the basis that the sooner deforestation is slowed, the more carbon in aggregate will be retained in the biomass of these forests. For the two largest tropical rainforest nations – Brazil and Indonesia – this point is particularly relevant, but even for low-income rainforest nations where large scale deforestation has not yet occurred, history clearly suggests that it is only a matter of time before forests begin to be rapidly logged and cleared to make way from high value cash crops and tree crops. Therefore, as suggested above, the sooner interventions and investments directed at providing strong incentives for this encroachment pattern not to be repeated in these countries are in place, the more likely it is that these measures will succeed. It should be noted here that the Democratic Republic of Congo – the third largest tropical rainforest nation – already has some large scale logging and conversion proposals under consideration.

There is also a larger global reality in play here. It seems likely that in international trade terms, the potential value of the REDD option will decline for every year that rainforest nations wishing to participate in it spend to become REDD-ready. Over the longer term investors will move to other options, and emitters will develop direct solutions (or other offset alternatives). It could be argued that the principal potential comparative advantage of the REDD process would arise if it were able to deliver significant carbon offset trades into the market very soon after international agreements on emissions reduction come into force: as we have argued above, there is unlikely to be a significant "crowding out" issue in those early years – in fact, the difficulty for large emitters unable in the short term to reduce their emissions directly is much more likely to be to find offset sources capable of delivering significant trades approvable under the post-Kyoto rules.

Under these circumstances, it would be particularly advantageous for the REDD approach if there were early rainforest country volunteers ready and willing to embark upon avoided deforestation programs. The underlying issue in this is risk: If rainforest countries wait until the post-Kyoto rules are in place before they begin action on the tasks outlined above, then they will certainly not enter the global markets for whichever global public goods they wish to sell until well after 2012. Some may not venture upon this transformation even then, simply because of the cost of bringing their market-readiness into place.

7.5.1 *An Emergency Package for Tropical Forests*

One initiative for early implementation has been proposed by The Prince's Rainforest Project[14] (PRP). The details of the proposal (Prince's Rainforest Project 2009 op cit) were presented by His Royal Highness, The Prince of Wales, to a meeting of national leaders and heads of international agencies in London, in April 2009, and by all accounts received an enthusiastic response.

The central idea of the proposal is to find a means by which to raise funds sufficient to provide effective incentives to rainforest nations to reduce their rates of deforestation. Unlike the standard REDD model, the proposal focuses on payment to rainforest nations for the potential benefits of all forest goods and services

---

[14] The Prince's Rainforests Project was established by the Prince of Wales in 2007, with the aims of encouraging consensus on how the rate of tropical deforestation might be slowed, and to design a mechanism by which rainforest nations could be compensated for reducing deforestation.

produced from reduced deforestation, rather than simply the immediately marketable one of forest carbon. In effect, the proposal sees the carbon good as an enabler of financial exchange based on the full suite of ecological, social and economic benefits that accrue to sustainable management of the rainforests, rather than viewing forest carbon as the prime motivating factor.

This frees the initiative from the difficulties of establishment and operation of a market for forest carbon; the financial transfers to a given participating rainforest country will be determined essentially by negotiation of the price of the incentive needed to persuade all involved parties to cooperate in the deforestation reduction effort, with payments to be made as necessary to supplement existing donor agencies efforts to build capacity for REDD readiness, and then further payments of the incentive funding on a delivered deforestation reduction basis.

The obvious question which arises is: how could this approach be financed, given that it will not rely on carbon market proceeds? The project team looked at the public funding options that might supplement existing donor agency funding commitments in this area:

- Hypothecation of taxes to this purpose.
- Directing surcharges applied to carbon emission-creating products (e.g. oil and petroleum products), or to non life insurance policies (representing a climate change premium) or to other businesses with links to the rainforests (pharmaceutical companies).
- Directing a proportion of funds raised through auction of emission permits.

While none of these should be rejected, we need to recognize that in the forthcoming emissions reduction economic scenario, competition in Annex I countries for public funds from the plethora of initiatives and projects aimed at emissions reduction is likely to be very strong. It is probable that such countries will raise their commitments to deforestation reduction initiatives when this appears feasible, but the aggregate amounts likely to come forth under the most optimistic assumptions about this source will fall well short of the amounts needed.

The Prince's Rainforests Project has come up with an interesting and elegant financing alternative, which would create a framework for private sector investment in reducing deforestation in tropical forests based on participation from large investors such as pension funds, insurance companies, mutual funds, national wealth funds and high net worth individuals. It would also involve the governments of Annex 1 countries in the financing, not just through partnership financing of capacity building for REDD readiness, but also through underwriting of a bond instrument which would provide the necessary incentive financing.

The financial and institutional details of this approach are complicated, and are still in a state of development as this book is being written, and readers are referred to the paper presented to leaders of the G20 group (Prince's Rainforest Project 2009 op cit) for an extended explanation of the concept. For our purposes here, we will summarize some of the essentials from that PRP paper:

The Rainforest Bond: As we have noted in this book a number of times, the scale of the financing required to support a serious global programme of deforestation reduction in the rainforests of the world dwarfs anything that has been available on a grant funding basis from Annex 1 country donor agencies for this purpose. For this reason, it is envisaged that a Rainforest Bond (in fact a series of them) would be raised in global capital markets. The bond markets represent enormous pools of liquidity; the PRP paper points out that governments and government backed entities in 2008 issued over US $3 trillion in bonds. A Rainforest Bond would be very similar in concept to what are known as Sovereign, Supranational and Agency Bonds, which are used by government treasuries, the World Bank and European Investment Bank and others. The market for this category of bonds in Euros and US dollars alone was US $400 billion in 2008, and so the issue of Rainforest Bonds of, say, US $10 billion per annum would be easily absorbed by these markets.

A precedent for this approach, cited in the PRP paper, is being implemented by the International Finance Facility for Immunisation (IFFIm), created in 2006 as a financing vehicle for the Global Alliance on Vaccines and Immunisation. With strong donor backing, the IFFIm has been able to gain a AAA credit risk rating from the three major ratings agencies. This has enabled the IFFIm to sell a US $1 billion bond in 2006, and the facility expects to raise a further US $4 billion over the next decade.

In the Rainforest Bond case, a term of at least 10 years would be used, to cover the likely implementation period for deforestation reduction programmes. The repayment schedule would be designed to suit the bond holders and the underwriters: most current bonds offer fixed annual interest payments to the holder, with payment of the principal at maturity, but it is possible to vary this so that all interest and principal is paid out at maturity, or, conversely, that a portion of the principal and the interest is payable each year.

Repaying the bond: In meetings with potential large institutional purchasers of a Rainforest Bond, the PRP Project has found strong interest in this bond issue, with its clear relationship to the global environmental issue of rainforests, provided that the bonds have the highest credit risk rating (AAA), and the yield is at least competitive with other AAA rated fixed income securities. However, the requirement for AAA-rating in this case would mean that the bonds will need to be underwritten by the Annex I countries, who already do provide this support for bond raisings made by the World Bank, and its private sector branch, The International Finance Corporation.

It would be desirable if ways could be found to reduce, or eliminate the liabilities of underwriting governments; obviously, the lower the liability Annex 1 countries are required to take on in this regard, the more likely it is they will support this early implementation option:

One approach which the PRP has considered in this regard would be for governments underwriting the programme to negotiate sharing of REDD payments, when these become available following the negotiation process set up after the Copenhagen UNFCCC meeting in 2009, with rainforest nations which have undertaken deforestation reduction under the Emergency Package early implementation activities. Obviously, Annex 1 countries cannot have any claim on REDD funds that are generated in this regard, especially since these will be generated by emissions offsets from these same countries. They might, however, have some claim to *additional* REDD revenues that have accrued to rainforest nations because of assistance they have received under the early implementation programme. On balance, the PRP believes that rainforest nations will probably be most unwilling to share a future revenue stream in this way, and therefore assesses this option as unlikely to succeed.

Another approach would be for rainforest payments under the bond approach outlined above to be linked to a "green investment fund" that would invest in clean development projects that generate returns sufficient to cover its own capital costs and to cover the interest and principal on the Rainforest Bond as well. This would not only reduce (or eliminate) the bond underwriting liabilities of Annex 1 governments, but would contribute to the broader climate change mitigation effort as well. For this approach to work, more bond financing would be needed, so that the rainforest payments could be made, and additional funds for income generating green investment elsewhere would be available. The PRP is continuing to evaluate this option.

Issuing the bond, and implementing the programme: There will, of course need to be national and international institutional capacity in place to ensure that the funds, once raised, can be disbursed effectively towards reducing deforestation in participating countries. The PRP paper emphasises that the key implementation step will be for each participating rainforest nation to conduct its own assessment of what payments it would need to effectively pursue a low deforestation pathway. Then, that country would need to negotiate with an international organization and come to agreement on payments to be made. For the present, the PRP has called this organization the Tropical Forests Facility, and has suggested that it would need the capacity to evaluate proposals from different rainforest nations, and seek consistency across the range of proposals, but ultimately the principle will be agreement based on negotiation in each case.

The PRP paper points out that one option for the agency to issue the bond would be the World Bank, which could issue Rainforest Bonds using its own balance sheet and AAA rating. This would have the advantage of utilizing existing structures and the Bank's credit rating. However, the use of funds would then have to follow the rules and procedures of the Bank, and this could be constraining on this new and untried initiative. The alternative would be for the Tropical Forests Facility to be established in a manner which would allow it to issue bonds itself, which is basically the approach used by the Global Alliance on Vaccines and Immunisation, outlined above. This would certainly be more flexible, but also may be more expensive, because the new facility would have a weaker balance sheet. The PRP does not express a preference at this stage for which option should be chosen, but suggests instead that more analysis will be needed to decide this question.

The PRP has consulted widely with rainforest nations, to determine their willingness and readiness to engage in the process, and expresses confidence that a large number of countries are ready to engage now, or could be helped through a preparatory process within 1 or 2 years, to become ready.

The PRP emphasizes that the Emergency Package proposal is intended as a short term, interim approach; an implementation period of 5–10 years is suggested. Thus, according to the PRP, it should not be considered as an alternative to or substitute for REDD or, for that matter, any other deforestation reduction approach that might be developed under the UNFCCC arrangements.

In our view, the Emergency Package would offer a useful testing ground, under real-world conditions rather than hothouse pilot studies, for testing monitoring systems, developing baselines, and other technical, institutional and policy options which are going to be vital to the success of REDD in the longer term. Drawing an even longer bow, we can see the possibility that if the Emergency Package approach worked up to or beyond expectations, then its modus operandi could well be picked up by REDD as negotiations proceed. If it did function well at national scale, then it might provide a viable alternative to the development of a REDD forest carbon offset trading scheme, with all the attendant technical and political issues and constraints outlined above in this chapter that that approach will bring.

References

Barr C (1999) Banking on sustainability: a critical assessment of structural adjustment in indonesia's forest and estate crops industries. CIFOR-WWF, Bogor, Indonesia

Benhin J, Barbier E (2000) Estimating the biodiversity effects of structural adjustment in Ghana. In: Perrings C (ed) The economics of biodiversity conservation in Sub-Saharan Africa. Edward Elgar, London, pp 268–308

Blaser, J., & Robledo, C. (2007). Initial Analysis on the Mitigation Potential in the Forestry Sector. Prepared for the UNFCCC Secretariat

Bockstael N, Freeman A, Kopp R, Portney P, Smith V (2000) On measuring economic values for nature. Environ Sci Technol 34(8):1384–1389

Braat L, Ten Brink P (eds) (2008) The cost of policy inaction: the cost of not meeting the 2010 biodiversity target. Alterra, Wageningen/Brussels

Butler RA, Koh LP, Ghazoul J (2009) REDD in the red: palm oil could undermine carbon payment schemes. Conserv Lett J Soc Conserv Biol. doi:10.1111/j.1755-263X.2009.00047.x

CIFOR, IPAM and ODI (2008) Integrating REDD into the global climate protection regime: proposals and implications. Background paper for the introductory meeting for the collaborative analysis. Tokyo, 24th June 2008

McKinsey & Company (2009) Global GHG Abatement Cost Curve v2. ClimateWorks Foundation/ McKinsey & Company: Project Catalyst

Corley RHV (2008) How much oil palm do we need? Environ Sci Policy. doi:10.1016/j.envsci2008.10.011

Costanza R, D'Arge R, de Groot R, Forbes S, Grosso M, Hannon B, Limburg K, Naeem S, O'Neil RV, Ruskin RG, Sutton P, van dem Belt (1997) The value of the world's ecosystem services and natural capital. Nature 387:253

Daviet F, McMahon H, Bradley R, Stolle F, Nakhooda S (2008) REDD Flags: What We need to know about the options, world resources institute posting on Poverty Environment Net, May 2008

Dixon S (1995) Introduction: the nature of structural adjustment. In: Simon, Spendgen, Dixon, Naerman, (eds) Structurally adjusted Africa: poverty, debt and basic needs. London/Boulder: Pluto Press

Eliasch J et al (2008) Climate change: financing the forests. Independent report to the Government of the United Kingdom, commissioned by the Prime Minister

Fairhurst T, McLaughlin D (2009) Sustainable oil palm development on degraded land in Kalimantan. World Wildlife Fund, Washington, DC

Glover D (1995) Structural adjustment and the environment. J Int Dev 7:285–289

Grieg-Gran M (2006) The cost of avoiding deforestation. International Institute for Environment and Development, London

Gueorguieva A, Bolt K (2003) A critical review of the literature on structural adjustment and the environment. Environmental Economics Series. Environment Department Papers No. 90. World Bank, Washington, DC

Houghton RA (2005) Tropical deforesation as a source of greenhouse gas emission. In: Moutinho P, Schwartzman S (eds) Tropical deforestation and climate change. IPAM/Environmental Defense, Belem

IPCC (2007) WGI Chapter 7, 2007, Intergovernmental Panel on Climate Change

Keith H, Mackey B, Lindenmayer D (2009) Re-evaluation of forest biomass carbon stocks and lessons from the world's most carbon dense forests. Proc Natl Acad Sci USA July 14, 2009 106(28)

Lewis S et al. (2008) Letter to Nature

Luyssaert S et al. (2008) Old growth forests as global carbon sinks. Letter to Nature 455, 11 September

Mainhardt H (2001) IMF intervention in indonesia: undermining macroeconomic stability and sustainable development by perpetuating deforestation. WWF-Macroeconomics for Sustainable Development Program Office, Washington, DC

Munasinghe M, Cruz W (1994) Economywide policies and the environment: lessons from experience. World Bank Environment Paper No. 10, Washington, DC

Oberthur S, Ott HE (1999) The Kyoto protocol: international climate policy for the 21st century. Springer, Heidelberg

Pandey K, Wheeler D (2001) Structural adjustment and forest resources: the impact of World Bank operations. World Bank Development Research Group, World Bank, Washington, DC

Panayotou T, Hupe K (1996) Environmental impacts of structural adjustment programmes: synthesis and recommendations. Environmental Economics Series. Paper No. 21. United Nations Environment Programme, Environment and Economics Unit

Parker C, Mitchell A, Trivedi M, Mardas N, Sosis K (2008) The Little REDD+ Book. Global Canopy Programme, Oxford, UK

Pearce DW (1998) Auditing the earth. Environment 40(2): 23–28

Pearce DW (2001) The economic value of forest ecosystems. Ecosys Health 7, December 2001

Persson A, Munasinghe M (1995) Natural resource management and economywide policies in costa rica: a Computable General Equilibrium (CGE) modeling approach. World Bank Econ Rev 9(2):259–285

Prince's Rainforest Project (2009) An emergency package for tropical forests. The Prince's Rainforests Project, Clarence House, London

Prince's Rainforests Project (2008) A plan for emergency funding: consultative document. Annexes. The Prince's Rainforests Project, Clarence House, London

Putz FE, Zuidema PA, PinardMA, Boot RGA, Sayer JA, Sheil, D., Sist, P., Elias, Vanclay, J.K. (2008). Improved Tropical Forest Management for Carbon Retention. *PLos Biol* 6(7): e166 dol:10.1371/journal.pbio.0060166

Reed D (1992) Structural adjustment and the environment. Westview, Boulder

Repetto R, Cruz W (1992) The environmental effects of stabilization and structural adjustment programs: the Philippines case study. World Resources Institute, Washington, DC

Seymour F, Dubash N (2000) The right conditions: the World Bank, structural adjustment and forest policy reform. World Resources Institute, Washington, DC

Shandra JM, Shor E, Maynard G, London B (2008) Debt, structural adjustment and deforestation. 1 1:1–21

Sheil, D. Casson, A. Meijard, E. van Noordwijk, M. Gaskell, J. Wertz, K. & Kanninen, M. (2009) The impacts and opportunities of oil palm in Southeast Asia: what do we know and what do we need to know? *CIFOR Occasional Paper*, Center for International Forestry Research, Bogor, Indonesia

Stern NH (2006) Stern review: the economics of climate change. HM Treasury, Government of the United Kingdom, UK

References

USDE (2005) Genomics: GTL roadmap; systems biology for energy and environment, U.S. Department of Energy Office of Science, August 2005. DOE/SC-0090

Warford J, Schwab A, Cruz W, Hansen S (1994) The evolution of environmental concerns in adjustment lending: a review. Environment Department Working Paper No. 65. World Bank, Washington, DC

White A, Martin A (2002) Who owns the world's forests? Forest tenure and forests in transition. Forest Trends and Center for Environmental Law, Washington, DC

World Bank (2005) Development Policy Lending and Forest Outcomes, Report No. 33537-GLB

World Wildlife Fund (2008) Policy approaches and positive incentives for reducing emissions from deforestation and degradation (REDD). WWF Discussion Paper

World Wildlife Fund (1994) Structural adjustment and the environment: Jamaica country study. Washington, DC

Wunder S (2003) Oil wealth and the fate of the forest: a comparative study of eight tropical countries. Routledge, New York

Wunder S, Sunderlin W (2004) Oil, macroeconomics and forests: assessing the linkages. The World Bank Res Observ. 19(2), World Bank

Young C, Bishop J (1995) Adjustment policies and the environment: a critical review of the literature. Working Paper Series 1, Center for Research in Experimental Economics and Political Decision Making (CREED), Amsterdam

Chapter 8
Final Thoughts

Abstract This chapter draws some overarching observations from what has gone before in this book, framed in the context of a more general paradigm linking social, cultural and environmental change. We follow some new work on these large issues which suggests that creating institutions to meet the challenge of sustainability is the most critical task confronting society; these institutions will need to be capable of integrating the various ecological, economic and social disciplines in a way that will allow continual adaptation to cycles of growth, accumulation, restructuring and renewal.

We suggest that in this new era, the primary reason for slowing global deforestation is that without this, there will be little chance of bringing anthropogenic emissions of greenhouse gases down to the levels needed to avoid catastrophic climate change. Dealing effectively with life in the Anthropocene is going to involve assigning high global priority to getting this done – especially for the tropical rainforests much more effectively than we have managed to do so far. In this sense, the forest carbon product – combined with a broader set of forest ecosystem values – represents a new opportunity; one which has the potential to fundamentally re-order the economic priorities of forest management and use at a global scale. We will need to see this for what it is – a market opportunity.

The crux of what needs to be done is to more effectively link finance and capital – those creations of the last renaissance which have powered both the best and the worst of our historical trajectory since then – to the natural forest systems, and ongoing human interactions with those systems, to shift the dynamic towards sustainability. Inevitably, we believe, this will require a much larger role for international private sector investment than has hitherto been the case.

This approach may raise concerns in some quarters, related to the role of international donor assistance institutions, the perceived need to protect biodiversity as a goal in itself, and addressing equity and rights issues as a priority. We acknowledge these concerns, but we maintain the argument that has underlain much of what has been written in this book: more compromise, and changes in approach are going to be necessary. Many of the criticisms we have raised of historical efforts of the international forests constituency to implement forest sustainability and reduced deforestation in the past – inconsistency and lack of coordination, inadequate financing, and the pursuit of ideological and unrealistic solutions – will remain as constraints under a new approach to this problem, unless there is new resolve on the part of all significant players to not allow the perfect to continue to get in the way of the good.

Books like this one commonly end with a collection of conclusions and recommendations, including – as often as not – an attempt at some institutional and policy architecture which will bring about the desired solution. We have spent a considerable amount of time in our involvement with international forestry visiting the various cemeteries and remains of such things – sometimes running across creations of our own which have suffered this fate, and sometimes, alarmingly, encountering reincarnations of programmes we had imagined to be long defunct. So we are not inclined to follow this route; the reader who has persisted to this point will be in little doubt as to our impressions and opinions on the issues and matters we have addressed in this book, and hopefully will have some idea of what we are suggesting needs to be done, and what needs not to be done.

We will re-visit some of our broad impressions in what follows, but we would like to do so in the context of what we think is the central intellectual task that needs to be undertaken in international forestry; to engage with the issue at the level of social and cultural change, linked to the broader issue of global environmental sustainability. To do that, we will first take a brief sojourn into some current thinking on the new paradigm that humanity needs to put in place, in order to really be capable of addressing the major environmental and climatic dilemmas we now face. We will then attempt to link this to the forests task.

8.1 The Search for a New Paradigm

In an essay published in the June 6–7, 2009 edition of an Australian newspaper, *The Sydney Morning Herald*, James Thornton[1] argues that the entirety of human society and culture has evolved in the 12,000 years of the Holocene era, with its benign climate and abundant natural resources. Now, however, a turning point has been reached; we have arrived at the beginning of the Anthropocene era – a human-made age which, as a result of actions we have perpetrated, and continue to do, is likely to be a lot less hospitable and generous to the continued rise of humanity. Since the advent of the industrial revolution, humanity has acted, in Thornton's language, as gods, though unconsciously. We have, he suggests, become capable of changing the seas, the atmosphere, the great forests – but the changes we have wrought are not to our long term advantage, nor without potentially serious consequences for ourselves.

---

[1] Thornton is a lawyer, and is CEO of a non-government organization named ClientEarth, which works to protect the environment through advocacy, lobbying, litigation and research. More articles by Thornton can be found on the ClientEarth website, http://www.clientearth.com.

To deal with this, Thornton suggests we need to adopt a much broader and long term approach to social and cultural change. He cites a collection of studies edited by Gunderson and Holling (2001), in which the case is argued for a paradigm shift in understanding how human and natural systems are linked. These authors have suggested that creating institutions to meet the challenge of sustainability is the most critical task confronting society; these institutions will need to be capable of integrating the various ecological, economic and social disciplines in a way that will allow continual adaptation to cycles of growth, accumulation, restructuring and renewal.

Thornton uses the forest as a natural world analogy of what is being suggested here: At the smallest and most short-lived end of the scale are the leaves, which serve a vital and immediate role in the life of the tree, and then fall to the ground, to be recycled. Next up the time scale is the tree: it may live for decades, centuries, or even millennia, absorbing and transpiring huge quantities of water in that time; producing all that leaf litter; maintaining the environment as a forest. The forest itself, may cover thousands of square kilometres, and survive as an ecosystem for hundreds of thousands of years.

Translating this to the topic of social and cultural change in an uncertain environment, Thornton argues that advocates of social change are focusing on the leaves: they react to the immediate event or harm, instead of trying to shape a narrative that can drive events. It is the narrative of the culture that controls events, not the next bill in the legislature: Thornton refers to this as the long, slow variable. Until our shared narrative articulates the world as we wish it to be, then this view of what is needed will never become reality.

Some readers will recoil at the vaulting ambition contained within these arguments, seeing them as impractical hubris. Others will argue that this limiting scepticism is itself a manifestation of the end of a social and cultural era; one which has reduced large segments of society to a dominating focus on profit and pleasure, in some ways reminiscent of the constraining of human potential that results when fundamentalism of the religious or political sort take hold.

Thornton points out that humanity has come through a major turning point at least once before, as the Renaissance of the fifteenth century re-arranged the historical narrative comprehensively: many of the religious strictures of the Middle Ages were thrown off; and a burgeoning of new art, politics, and the advent of the new concepts of banking and finance began. Yet some of this change was founded on old ideas – in that case the classical texts of Greek and Roman scholars, which were used to develop the concept of humanism.

8.2 What Does All This Mean for the Forests?

One thing we can be sure of is that, under the above interpretation, things will not be stable and predictable. Our habitual reluctance to face major change, and to prefer chasing small victories, planting ideological flags on programmes and initiatives in the field, and marketing a futile gradualism in place of fundamental change, will not address the situation. Indeed, all of these tendencies, which we have seen so much of in the global forestry world, can be likened to gardening in the pathway of an oncoming wildfire. They must give way to more unified and much more ambitious transformations.

One of those, we suggest, is that deforestation must be slowed significantly if there is to be any chance of bringing anthropogenic emissions of greenhouse gases down to the levels needed to avoid catastrophic climate change. Dealing effectively with life in the Anthropocene is going to involve assigning high global priority to getting this done – especially for the tropical rainforests – much more effectively than we have managed to do so far.

We acknowledge that this view imparts a very specific priority to the climate issue, in relation to forest outcomes, and in doing so may be seen by some as minimizing the broader concerns that many have about forest loss: the continuing risk to biodiversity, and larger natural resource equilibrium issues, including rainfall, land stability and soil quality; the importance of forests in poverty alleviation and economic security, because of the nature of human settlement in and around forest areas, especially in the tropics; and other concerns. Our intent here is not to minimize or trivialize the many and various uses and values of natural forests, but to set some possibility boundaries about what can be done with the new opportunity that carbon forestry raises: to ignore that, in our view, is to lose touch with Thornton's long, slow variable.

We probably do not need to remind readers of the fact that concerns with the broad suite of forest issues on the part of the international forests constituency[2] have not been successfully translated into effective action at scale in those forests. The forest carbon product (possibly, as noted in Chapter 7, combined with a broader set of forest ecosystem values) represents a new opportunity; one which has the potential to fundamentally re-order the economic priorities of forest management and use at a global scale: we need to see this for what it is – a market opportunity – and we need to refrain from attempting to load all sorts of hitherto failed earlier goals onto it. If we can contain ourselves in this way, then broader possibilities will open up for forests in the future, as the recognition that maintaining forest cover is not an option, but is a necessity for humanity, begins to take hold.

This is the paradigm shift we seek; the long, slow variable; the one that is capable of eventually opening up all those other possibilities for the global forests. It will not be created by conceptualizing majestic new visions and ideologies, but rather by focusing on the specific new opportunity which has global reach and scale first, and then allowing the attitudinal and cultural shifts to build from that.

In the opening two chapters of this book, we have argued that the atmosphere of conflict and the ever-present permanent campaign mentality in the debates and controversies that have raged in the international forests constituency has in all

---

[2] Very broadly defined, in this case, to include the many concerned members of the public who have supported forest conservation and protection programmes and initiatives which have been launched in the past two or three decades.

probability been a constraint on the entry of "green capital" into the sector from global capital markets[3]. What we will need are some significant successes – real and significant ones, not confections for the purpose of public relations benefits. Nothing will drive the impact on the deep cultural and political changes that will be needed in the broader public mind more than some clear and sustainable reductions in deforestation; it will encourage investors and resource owners alike to persist with the strategy, and the visibility of success will permeate as this proceeds.

The crux of what needs to be done to achieve this is to more effectively link finance and capital – those creations of the last renaissance which have powered both the best and the worst of our historical trajectory since then – to the natural forest systems, and ongoing human interactions with those systems, to shift the dynamic towards sustainability. Obviously, from what we have said so far, we would not want this to be interpreted as support for the idea of allowing " a thousand blossoms to bloom" in the new paradigm for forests: As we have tried to make clear throughout this book, the lack of discipline and cohesion in advocacy for sustaining forests, and the failure to apply a rational hierarchy to measures and initiatives adopted in the name of deforestation reduction, have contributed significantly to the lack of success in what has been attempted to date, and could well do so again, as new options, possibilities and threats open up as the Anthropocene era takes hold.

Given all this, we need finally to attempt to enunciate some general principles arising from the issues and arguments we have examined in this book, which, if put into practice, would raise the prospects of sustaining forests and changing the background culture that affects these forests.

8.2.1 Formulating the Link Between Forests and Capital

As we have reported from many studies on this subject in this book, large amounts of capital are going to be needed to finance global forests sustainability effectively, and the goods and services that will be produced as a result, must be the priority concern for all involved in the forests. We have argued that neither international donor funding, nor national macroeconomic reforms and investments by rainforest nations by themselves will provide effective levels of funding for forest sustainability.

The international private sector will therefore have to be the source of much of the funding and, like it or not, this will apply certain constraints and directions on what is supported. At present, obviously, the primary avenue for this is the sequestered forest carbon market, for which, potentially, there is strong international demand. Our argument is that this is not yet the case for many of the other goods and services from forests – the biodiversity, soil, water and rainfall values – although as we have argued in this book, in due course this could change as the global consequences of natural resource depletion beyond the immediately evident climate impact become more apparent.

---

[3] Except perhaps for purposes of green-washing corporate activities (which, in our view, does not really count, if we are in the paradigm-shifting business).

Until this occurs, however, it would be better from a strategic point of view to regard these other goods and services as ancillary benefits of an investment in forest sustainability. This might at first appear simple; it might seem that the idea of maintaining high carbon loads on a given piece of forest is fully compatible with high level biodiversity protection or other non-invasive use, and in biological terms this is true. However, we need to bear in mind here that in financial and market terms, the forests of primary interest in the forest carbon market are those which are presently, or potentially, at high risk of removal or degradation. This is obvious: unless the forests are in this situation, they are of little interest to purchasers of forest carbon produced (for market purposes) by retaining forest cover at risk of removal. We have suggested in this book that these will mainly be forests which are commercially accessible for logging, and/or attractive for conversion to other forms of land use. By and large, we should be prepared to consider that maintaining these forests under sustainable management, which would in most cases include the logging option (applied in as low-impact and sustainable a manner as possible, since maximizing of retained carbon will depend upon this), would in most cases be a good result within the market and financial constraints that exist. Attempting to import into this situation agendas relating to high level biodiversity protection and other very low-intensity uses of such forests could run the risk of becoming a major disincentive to the carbon forestry option.

We should also recognize that this constraint will apply to the minority element in financing sustainability that is comprised of donor funding: there is always an opportunity cost to directing such funds towards any specific part of the suite of forest outcomes, no matter how much this may be desired by some interest groups. For example, taking funds away from capacity building for REDD-readiness in rainforest nations (which by definition will be focused on the most commercially attractive forested sites) to undertake activities that are not related to this task could slow the progress of financing reduced deforestation, and therefore the rate of financial flows from the largest potential source into the sector in aggregate: the test of an alternative activity should, in our view, therefore be the rigorous one of assessing the extent to which this is likely to happen. There are symmetries and dependencies here between the various sources of financing of reduced deforestation available that must be considered when plans for disbursement are made.

8.2.2 *Balancing Equity and Effectiveness*

We have acknowledged in various places in this book that the idea of extending equity in forested areas to local communities, indigenous groups and other social groups who, like these, have in many cases been excluded from the benefits of forest utilization, is a morally worthy goal. Moreover, in many cases it will be a necessary approach to reducing conflict and building effectiveness in forest management – outcomes that will be indispensible to successful sustainable forest management over the large areas of forest where community rights and participation are important concerns.

However, as with any other consideration, there will be limits and constraints on this. We have suggested, for example, that there will in many cases be a natural tension between pursuing actual titling of forested areas to these groups, and the rapid implementation of sustainable management for carbon forestry and related purposes. Experience shows that land titling, even under the most amenable of circumstances, is a long term process – and because of the frontier nature of forests, characterized in many places by the influx of new immigrant groups, and innate conflicts between different groups and communities present, these areas will be most difficult to deal with in this way. The potential for conflict, while the titling process is in progress, may actually be higher, rather than lower, than the background level of conflict between interest groups that occurs now in such areas, and this would certainly act as a disincentive for any potential investor in carbon forestry, especially large international investors who will have many reasons related to political, financial and reputational risk to avoid such situations.

Our view is that interim measures, such as extending legally sound rights of access and usage of forest areas to local groups and communities, and possibly involving them in predetermined shares of the carbon rights to forest areas under consideration for intensive sustainable management, will in general be compatible with effective sustainable forest management. Obviously, the specific solutions adopted will vary from one location to another, and any attempt to impose boilerplate rules under a global REDD agreement relating to the means by which local communities are engaged in the process on negotiators of a locally specific agreement should be rejected. The immediate criterion of success in this regard should be significantly improved involvement of local communities and groups in the forest carbon programme; the longer term goal should be a progressive improvement in the livelihood and economic security opportunities for local communities, as a result of the benefits generated by the carbon forestry programme itself.

8.2.3 Keeping Watch Over the Market Instrument

In this book, we have examined the international market for forest products; the markets for products that are seen to compete with natural forests for land – oil palm, soybean and the like; and some market issues related to REDD.

Forest products: Trade in forest products has certainly been associated with forest destruction in the past, but we have cautioned that this does not mean that application of trade specific sanctions or even current means of certifying that trade, are necessarily the best answers to this. The key to profitable and sustainable trade in forest products is a clear understanding between buyers and sellers as to the rules and objectives of that trade, rather than an attempt to apply leverage in the market by one side or the other.

In Chapter 4 of this book, we have argued that, in general, there should be a multilateral solution to the problem of trade regulation rather than relying on importing country or bilateral arrangements to facilitate implementation to companies which supply different international markets. In addition, it should be ensured that these instruments do not become disguised measures of discrimination, a concern already raised about forest certification, which is has become a de facto prerequisite in access to some markets.

We have also briefly examined in this book the Voluntary Partnership Agreement, introduced under the EU Forest Law Enforcement, Governance and Trade (FLEGT) programme. This agreement would require acceptable proof that logs, sawnwood and plywood entering the EU market that these products come from legal sources. But, again, VPAs will have limitations without a broader multilateral participation which could address trade diversion and a holistic approach to dealing with the underlying incentives and drivers of illegal logging. Measures such as these will put pressure on importers, who will be liable if their products are not legally harvested, and this will impact upon the access of tropical timber in particular to main import markets. The problem is that none of these regulatory measures effectively addresses the underlying reasons for illegal logging and trade, nor the root causes of deforestation either. As will be seen immediately below, this problem of potentially counterproductive trade measures in relation to forests is not restricted to forest products markets alone.

Cash crops on forested land: In the case of the heavily traded cash crops and other activities such as grazing which have replaced large areas of tropical forests, we have pointed out the importance of understanding the process of forest degradation and deforestation, rather than simply identifying the end point – the advent of a new land use over a previously forested area – and then assuming this to be the primary cause.

This is an area where local nuance is important: in some cases, the new crop or other land use may have been the primary instrument of deforestation; in others, it may simply have been the last step in a long process with complicated social, cultural and political origins. This is one reason why the application of trade sanctions, or even attempts to generate a premium, or a consumer preference, in an end market for these products, may be inappropriate, or at least ineffective for the task of reducing pressure on forests. Another reason why the trade instrument needs to be carefully wielded in this situation is that, if that is all that is done, the result may well be alienation of the government and other parties in the supplier country, who will see this measure as a means of passing responsibility for emissions reduction to poorer countries. The essential point here is that measures to restrict trade, or to differentiate it in the market to the same effect, should only be applied in cooperation with supplier country interests, which could, for example, be obtained via financial assistance to transform production of these crops from unsustainable to sustainable methods. The need to integrate these approaches with the linkage of capital to forests, discussed earlier, will be self-evident here.

The REDD market: We have noted in this book the possibility that some governments, with the support of some environmental groups, may attempt to reduce the access of forest carbon offset trades arising from reduced natural forest deforestation in their markets, due to a fear of market flooding with these trades reducing incentive for high emitters to invest in direct emission reduction measures in country.

8.2 What Does All This Mean for the Forests?

There may be elements of protectionism in this, given that some developed countries have invested heavily in domestic emission reduction technologies. Alternatively, it may indicate a discouraging lack of commitment in those countries to significant emission reduction targets, meaning that the overall market intended will be small. Whether or not these play some role in the concern being manifested, in our view the threat of a market flooding development is slight: deforestation reduction in the major rainforest countries of the world will take place progressively over a number of years – perhaps ten or so – not all at once. And, when that high level of aggregate reduction is reached, the availability of reduced deforestation trades will of course decline, leaving plenty of room in the market for alternative emission reduction strategies. It would be better to regard the forest carbon option as a felicitous interim opportunity to mobilize emissions reduction relatively quickly, rather than as a threat to other options for reducing emissions.

Our search for a new paradigm for forest sustainability would certainly suffer if significant restrictions on the REDD market are seriously in prospect at this stage, before even the basic agreements on REDD have been fully negotiated. We suggest that the true comparative advantage of REDD trades will be their ability to buy some time until large emitter industries, such as coal-fired power stations, can overcome their capital fixity and invest in effective technologies to reduce emissions through their standard processes – or in low emission alternatives.

The necessity to build momentum in this way is evident: Certain countries (Australia being one) are contemplating not only limiting their commitments to extremely low near term emissions reduction targets under the forthcoming emissions reduction scheme there, but also issuing generous quotas of emission permits at no cost to major polluters – despite the lessons that should have been learned from Europe's experience with this approach. Annex I countries attempting to minimize their commitments are doing so precisely because they recognize the political difficulty of implementing direct emissions reductions in the short term, and the political consequences of failure. Consider how much better it would be if instead of this defeatist approach, such countries – in partnership with rainforest nations – were able to achieve some major emission reductions relatively early in the process, and set in train that fundamental transformation in thinking that is needed. Ultimately, of course, the high emitters would know that they must find another solution – not least because of the natural limitation on REDD trades over time, as discussed earlier – but at least they could participate, directly or indirectly, in a serious emission reduction effort in the meantime, under this option. This would also result in appropriate signals being transmitted to the renewable energy sectors in countries in this situation – a result that will certainly not emerge from insulating high emitters from any need to adapt in the medium term.

If reticence to open developed country markets to REDD to achieve this does become a reality, then the Prince's Rainforests Project alternative to REDD outlined at the end of the last chapter, which we have treated there primarily as an interim, early adoption possibility (given the fact that REDD will probably take a considerable time to negotiate and implement) would take on added significance. It would provide a means by which developed countries could underwrite investments in reduced deforestation, financed via a large bond raising, without the need to place reduced deforestation trades directly onto their markets. Instead, payment to rainforest nations for provision of the suite of all the forest goods and services, including forest carbon, would be via negotiated settlement in each country case, while access to the private sector finance market would be maintained via the bond raising.

8.2.4 International Debates on Forests: Impotency and Intransigence

A sub-text of much of what we have discussed in this chapter so far is the issue of potency and agency in the historical arrangements and dialogues which have been set up around the forest sector in recent years. Obviously, from what we have said so far, we would regard the REDD process as a *potentially* potent exchange: it has the backing of governments at the most senior level; it could be the pathway into a significant market for reduced deforestation, by way of the forest carbon good; and in this process it could tap significant new sources of finance from both the private and public sectors.

We would do well, however, to learn from the history of the global forests dialogue, before pronouncing victory in this particular case.

The main lesson is that this potential needs to be acted upon – and resulting programmes implemented – quickly. This is because of what we suggest is the iron rule of multilateral commitments to international development (in general, not just for forest issues): the longer the period of time that elapses following such a commitment before any implementation of *significant* activities in direct response to it is scheduled, the less potency the original commitment carries. As various participants in the original process notice that little is being done, they begin to withdraw from the substance of what was agreed, albeit (usually) while retaining the rhetoric in public pronouncements on the issue, until there is virtually no chance of any significant action occurring, at which point even the rhetoric is abandoned.

The intergovernmental dialogue on forests: In Chapter 6 of this book, we reviewed the intergovernmental dialogue on forests, which we suggest would be a case study exemplifying the above axiom on declining potency. As we saw, the official intergovernmental dialogue on forests has run now for almost 15 years, through its various incarnations as the Intergovernmental Panel on Forests, then the Intergovernmental Forum on Forests, to its present form as the United Nations Forum on Forests. The process has generated an enormous amount of technical and policy material, and many resolutions on issues it has considered.

At its sixth session, in 2006, the UNFF announced agreement on some major objectives on sustainable forest management, reducing forest cover loss, enhancing the benefits of forests, and increasing allocations of official development assistance to forests. However, these seem to be pretty much what the intergovernmental dialogue on forests has propounded for all of its existence, to little apparent effect. Regarding the last of these objectives – raising development assistance commitments to forests – it is true that these do appear to be rising at present. However, we would suggest that to attribute this to the UNFF process, rather than to larger matters related to climate change and REDD, would be rather like attributing the breaking of a drought to the last shaman or witchdoctor to perform the rainmaking rituals before a deluge occurred.

Perhaps we are being unfair here: in Chapter 6, we noted the fact that from its inception, the intergovernmental dialogue on forests was driven by deep differences on some basic policy issues between the governmental delegations which attended, and buffeted by criticisms from some of the NGO participants that the process was neither inclusive, nor effective. There was insufficient involvement at the senior political level to cut through these differences, and thus the force of whatever was agreed under the process was reduced to symbolic or ritualistic levels, requiring no real programme commitments. This suggests that, if nothing else, the design and concept of the dialogue was deeply flawed.

Intransigence and the weakest link: Our examination of the tribulations of the World Bank in regard to its approach and policy towards forests in Chapter 6 illustrates a tendency among some members of the international forests constituency towards grandstanding and intransigence, in place of effective action on their stated objectives in forests. In its normal core business, the World Bank is not an organization which lacks agency, or potency. However, it is fair to say that the Bank's participation in the environmental issue, from the time of the Rio Earth Summit, has at times taken the organization well out of its comfort zone. This was quickly recognized by some of the more ideologically inclined environmental NGOs, who then identified the Bank not as the worst offender in their sanctified world, but simply as the weakest link; the most attractive target. As we saw in Chapter 6, campaigning by these groups led managers in the Bank first to an appeasement solution, resulting in the policy of 1991, in which the Bank excluded itself from any involvement in production activities in tropical rainforests. This was seen by some of the NGOs as a great victory although, as was apparent at the time, there was no possibility that a single hectare of rainforest would be preserved as a result of adoption of this approach. Keeping the Bank out of investing in improved forest management helped no-one, except those who sought kudos from this result itself, rather than its field implications.

After persisting with this policy for the best part of 10 years, independent evaluators of Bank performance within the organization began to point out the illogicality of the Bank's position on forests, and then the Bank's board of executive directors asked for a re-appraisal of the policy. The resulting revision foreshadowed a cautious and highly proscribed rejection of the blanket ban of the earlier policy, but the prospect of even this departure from the doctrinaire ban re-invigorated the campaign juggernauts of the radical NGOs, as we described in Chapter 6.

The irony in all of this is that it happened just at the time when the option of financing sustainable forest management via the potential value of reduced deforestation carbon was becoming visible: the concept of allowing reduced deforestation to be used as an emissions offset on the emissions trading market had been mooted in discussions surrounding the first Kyoto Protocol agreement in 1997. As we have noted in Chapter 7, it had in fact been rejected in the initial negotiations around the Protocol, because at the time it was recognized that there had been too little preparation to include it, but it was firmly on the agenda for future inclusion, as we now know.

There is a great deal at stake for the forests in the advent of global interest in financing sustainability. Large programmes and ideas will be abroad; major financial resources will be in prospect; significant opportunities for members of the international forests constituency to ply their trade(s) to greater effect may be at hand. This will require that old rivalries and conflicts will need to put aside, and grand institutional and agency ambitions will need to be tempered, if the same dysfunctionality we have described is not to raise its head again.

This of course is not to suggest that no room should be made in the process for the many shades of opinion on how we should approach the issue of saving the world's remaining natural forests; it is right and proper that these opinions should be heard and debated. We cannot afford, however, for the situation to deteriorate into another bout of highly polarized campaigning; there is too much to lose from that, and the reality of getting done what we will need to get done will require maximum cooperation between all involved groups. This should not prove beyond the capacity of most groups involved; our failure so far as a constituency to bring the lofty rhetoric on saving the forests into some sort of reality leaves us with a sobering lesson on the costs of conflict in this task.

8.3 Some Final Words

Our purpose in re-visiting some old battlegrounds in this book has not been for historical interest, nor to set off new conflicts within the international forests constituency. It will be apparent to those who have been watching and participating in this business, as we have, that for every case study of dysfunctionality we have examined, there are ten – or in some cases, a hundred others we might have chosen. For every criticism we have made, there are many more we could have raised.

Our intention in this book has been to draw the lessons from the experience of international forestry over recent decades, and to attempt to apply these to the development of the new paradigm that is now needed. There are plenty of things that have gone wrong with individual programmes and projects in this forests world, and plenty that have gone right. Our focus, however, has been on the broader behavioural and ideological issues that have grown up around the processes involved, and the resulting dysfunctionalities that have plagued them at the institutional and programme level.

From the outset, we have made clear that we are not believers in the "silver bullet" approach to forests reform. We have been around for long enough to know that actually implementing effective deforestation reduction – in the difficult enabling environment of the tropical rainforests in particular – will be no picnic. A great deal of negotiation, analysis, institutional development, community engagement at large scale and then the implementation of very large financial deals to mobilize the effort in the field will be needed. REDD, or any of its possible variants and alternatives, will be a demanding taskmaster. Although, as we have suggested, it is technically capable of success within relatively short time periods, compared to other forms of emission reduction at a large scale, its implementation will need to be persisted with for some time, and even then success will not be forthcoming in some situations. The very last thing that is needed at this stage is a breakout of new campaigns and new breathless enthusiasms at the prospect of going forth and mobilizing all ideological creatures great and small – preceded, of course, by the usual unseemly competition for resources to underwrite these – instead of the dogged effort needed to complete negotiations, and implement the results in an efficient and well prioritized manner.

The rationale for a new paradigm, and urgent new action to implement it is clear, if humanity is to extricate itself from the colossal environmental dilemma it has created. The perfect storm aspect of the issue itself, as the confluence between climate change, long term global environmental unsustainability more generally, and the loss of rainforests flows ever more strongly through the consciousness of the global public, presents the international forests constituency – again (hopefully) very broadly defined in this case – with an opportunity (very possibly the last one) and an obligation to get it right. Forest sustainability can be seen as a very large pilot study of how to go about the overall task of greenhouse gas emissions globally: Because it is at least technically possible to bring reduced deforestation into the global marketplace relatively quickly – initially perhaps for the carbon resource, but ultimately for much more – it will provide us with an early look at the prospects of success.

We had better make that a good, hard look.

References

Gunderson CS, Holling L (2001) Panarchy: understanding transformations in human and natural systems Island Press, ISBN 1559638575, 9781559638579

Index

A
A case study: oil palm in Indonesia, 159
Agricultural technology and deforestation, 84
An emergency package for tropical forests, 187
Asking the wrong questions, 33
A tug-of-war between international schemes, 68

B
Biofuels, 88
Brazil, 13, 31, 45, 48, 61, 62, 68, 84–89, 91, 92, 95, 97, 101, 104, 118, 126, 129, 148, 157, 174, 186

C
Can trade rules differentiate sustainably produced forest products?, 62
Certification, 11, 68, 69, 71, 79, 124
CI. *See* Conservation International
Clean development mechanism (CDM), 42, 76, 173
Club of Rome, 28, 108, 109, 110, 112, 113, 145
Competitive Enterprise Institute, 28
Conservation International (CI), 135, 138–143
Convention on Biological Diversity (CBD), 63, 121

D
Defining forest, 43
Development banks, 4, 20, 21, 30, 125, 126, 130, 139, 167
Development policy lending, 137, 166, 169, 170, 193
Drivers of deforestation, 9, 13, 85, 96, 101, 102, 177
Dysfunction, 20

E
Ecological Gnosticism, 4
Ehrlich, P., 28, 110, 113
El Nino, 50

F
Fires, 4, 29, 50, 150
Food and Agriculture Organization (FAO), 9, 10, 14, 17, 19, 39, 43, 44–48, 52, 56, 58, 66, 67, 72, 79, 89, 123–126, 157
Forest and woodland area, and deforestation, 45
Forest carbon and production figures, 45
Forest certification, 62, 63, 67, 69, 70, 76, 78, 124, 201
Forest Law Enforcement and Governance (FLEG), 73, 79, 130
Forest Law Enforcement Governing Trade (FLEGT), 73
Forest policy in the World Bank, 130
Forests and climate change, 151
Forests and international conflict, 50
Forests and life, 50
Forests and rain, 49
Forests in the broader economy, 115
Free Trade Area of the Americas (FTAA), 65
FSC. *See* The Forest Stewardship Council

G
GATT. *See* General Agreement on Tariffs and Trade
General Agreement on Tariffs and Trade (GATT), 60, 63, 66
Global Canopy Programme, 49, 175, 192
Global forest cover and cover change, 44
Global Forest Resource Assessments, 17, 43, 52

Greenhouse gas (GHG), 21, 22, 29, 38, 112, 154, 176, 192, 206
Gus Speth, 113

H
Hardin, G., 82
Heartland Institute, 28

I
Illegal logging, 10, 36, 67, 71, 72, 74, 75, 96–98, 101, 103, 130, 161, 163, 201
Improving effectiveness of certification, 70
Indonesia, 13, 26, 33, 48, 49, 50, 62, 79, 84–86, 89–95, 97, 98, 104, 126, 127, 141, 144, 145, 157, 159–163, 165, 168, 169, 179, 186, 191, 192
Intergovernmental Forum on Forests, 6, 128, 203
International demand management, 98
International Finance Facility for Immunisation (IFFIm), 189
International Tropical Timber Organization (ITTO), 6, 59, 65, 72, 79, 122, 127
Is the world running out of wood?, 10

K
Krugman, P., 113
Kuznets curve, 83, 108, 111, 112, 114, 144, 179
Kyoto Protocol, 121, 153, 172, 173, 185, 192, 205

L
Lacey Act, 74, 75, 130
Landsat, 48
Lawrence Summers, 28, 108, 111
Leslie, A., 30

M
Malthus, 14, 28, 108, 113
Manicheans, 36, 38
Marrakech Agreement, 60
Millennium Development Goals, 7, 137
Mining, 94
MODIS, 48
Multilateral agreements on global environmental sustainability, 118

N
National Aeronautics and Space Administration (NASA), 49, 50, 154
New funds and mechanisms, 118
Non-legally binding instrument on all types of forests (NLBI), 64, 128, 144
Non-tariff barriers, 62, 168

O
Oil palm, 90, 91, 159, 160
Opportunity costs, 27, 34, 77, 91, 156, 157, 164, 180
Ostrom, E, 82

P
Payment for environmental services (PES), 76
Pearce, D., 149
Plurilateral Government Procurement Agreement, 63
Pre-human forest cover, 43
Programme for the Endorsement of Forest Certification (PEFC), 69

Q
Quantifying ecosystem values, 150

R
Reduced emissions from deforestation and forest degradation (REDD), 10, 17, 42, 76, 121, 164, 172, 174–189, 191, 192, 199, 200, 202, 206

S
Shifting cultivation, 95
Soy, 87
Stern, N., 29, 108
Sustainable forest management (SFM), 7, 12, 17, 30, 32, 35, 48, 62, 63, 67, 71, 78, 120, 123, 126, 128, 129, 133–136, 138, 139, 143, 148, 181, 200, 204, 205
Sustainable use of a common pool resource, 83

T
TBT Agreement, 60, 62
Technical Barriers to Trade. *See* TBT Agreement
TFAP. *See* The Tropical Forestry Action Plan

The Anthropocene, 196, 197, 198
The basics of REDD, 175
The Brundtland Report, 119, 122, 123
The "chilling effect" of Bank forests sector policy, 131
The Earth Summit, 120
The failure of forests sustainability, 148
The Food and Agriculture Organization (FAO), 9
The Forest Stewardship Council (FSC), 68, 69
The impact of burgeoning plantation and grazing commodities, 85
The intergovernmental dialogue on forests, 127
The new World Bank forests sector strategy and policy, 133
The problem you see depends on where you stand, 22
The Rainforest Bond, 188
The raw material base is also changing, 58
The right question, 35
The Rio Earth Summit, 120
The Stockholm Agreement, 119
The tragedy of the commons, 82
The Tropical Forestry Action Plan (TFAP), 124–126
The World Summit on Sustainable Development, 121
Trade has undergone major changes, 56
Trade-Related Intellectual Property Rights (TRIPS), 63
Tree plantations, 92–94
Tropical Forestry Action Programme, 6, 125
Tropical rainforests: a key concern, 48

U
UNFCCC. *See* United Nations Framework Convention on Climate Change
United Nations Forum on Forests (UNFF), 6, 64, 122, 123, 128, 203
United Nations Framework Convention on Climate Change (UNFCCC), 17, 42, 121, 152, 153, 157, 172, 173, 176, 178, 186, 189, 191

W
Westoby, J., 25, 26, 39
What have forests to do with global economics?, 28
Why another book on forests, 5
Winners and losers under trade liberalization, 61
Wood fuel, 95
World Bank, 6, 15, 28, 33, 71–73, 79, 98, 103, 108, 112, 118, 124, 125, 130, 132, 133, 135, 142, 144, 145, 167, 169, 170, 189, 190, 192, 193, 204
World Trade Organization (WTO), 60, 63, 122